Wine Management and Marketing 2

Series Editors
Jack Legrand and Gilles Trystram

Wine Management and Marketing 2

Responses of the Industry to Crises and New Expectations

Edited by

Foued Cheriet
Carole Maurel
Paul Amadieu
Hervé Hannin

iSTE

WILEY

First published 2024 in Great Britain and the United States by ISTE Ltd and John Wiley & Sons, Inc.

ISTE Ltd
27-37 St George's Road
London SW19 4EU
UK

www.iste.co.uk

John Wiley & Sons, Inc.
111 River Street
Hoboken, NJ 07030
USA

www.wiley.com

Library of Congress Control Number: 2024931119

British Library Cataloguing-in-Publication Data
A CIP record for this book is available from the British Library
ISBN 978-1-78630-874-0

Contents

Foreword

This is the first time I have been asked to write the foreword to a collective book like this, and I would first like to warmly thank the coordinators of this publication for taking this risk, because I admit that at first, I was a bit puzzled as to how to go about it. In view of the contents of the book, a common thread seemed possible to me for this foreword: to consider the sectoral specificity of the vine and wine sector, in order to address the subjects covered in this book.

The wine sector has some unique characteristics, on different levels, which are briefly mentioned as follows; the main ones being: wine, a product of a perennial plant, contains alcohol, is miscible and storable and, despite the significance of the geographical indication, is increasingly substitutable by another wine. These specificities induce particular modalities of market operation.

Thus to begin with, the vine has some essential characteristics, the first of which seems to me to be the length of its productive cycle. Moreover, shortening the wine-growing time (supply) to best match the wine making process time (demand), which is experiencing faster and sometimes unpredictable changes, has, since antiquity, been an important axis of public intervention in viticulture. Grubbing-up and restructuring premiums for the vineyard mainly pursue this objective. For the record, the first grubbing-up regulation in history, intended certainly to free up agricultural land for food production as well as to reduce competition with transalpine wines, was the edict of Domitian, reported by the Roman historian Suetonius, born around 70 AD, which provided for the grubbing-up of half of the vineyards of Gaul.

This search for long-term adjustment is all the more difficult because the productive cycle of the vine is very long, and thus managing an area planted with vines, especially in an area with a well-known geographical indication, is more like managing a productive investment. Depreciation in this case is close to an accounting fiction: no one knows the renewal value of a property on average 35 years later. Nevertheless, this characteristic leads to the search for an asset valuation that is currently "immobilized on the spot". This is certainly not uncommon in this sector to the historical success of differentiation by geographical indication, and in particular PDO (Protected Designation of Origin). It also leads to the search for valorization, for example, of wine-growing landscapes via wine tourism.

A second set of characteristics of the sector that is interesting to consider is that relating to the major product from vine cultivation in our countryside, namely wine. It is of course a liquid product containing alcohol and is, therefore, miscible and quite easily storable, even if these characteristics are especially true in red wine, more than in rosé or white, where, except for high-end products, transfer of stocks in satisfactory quality conditions hardly exceeds one year. Thus, a cyclical surplus becomes a structural surplus, if it is not managed in one way or another, i.e. it is at least "transported" to the next rural area.

This risk of market imbalance as well as the alcohol content of wine, and by-products of winemaking, have made distillation the traditionally-preferred method of managing wine surpluses in the European Union until very recently. Moreover, this alcohol content has a second consequence: faced with societal expectations such as respect for both human health and nature protection (*One Health* approach), CSR (Corporate Social Responsibility) of companies in the sector should be rather oriented by the search for a reduction of inputs, especially pesticides, in the near future. It is quite difficult to imagine, despite wine dealcoholizing methods, a sector totally free of ethanol and its potential consequences on public health, especially in the event of excess consumption.

Finally, it can also be interesting to consider some of the characteristics of the wine market and how it works. This concerns the importance, for entry-level customers in particular, of the large and medium-sized retail channels of general food distribution, and therefore the expectations of these operators in terms of logistics, on the domestic market as well as for export, frequently via importers. Although this characteristic of sectoral distribution is not exceptional, despite a trend towards local purchases and, in some vineyards, fairly substantial property sales, the trends towards the massification of purchases and the substitutability of provenances in export markets are underpinned by a few important vectors:

– In my opinion, the first of these would be increasing urbanization: in December 2022, according to the World Bank, 56% of the world's population, or 4.4 billion people, live in cities, and this trend is set to continue. Doubling of the current number of urban dwellers by 2050 could lead to almost seven out of ten people in the world living in urban areas.

– The second vector could be the likely approximation expected from bottling places to consumption places, to reduce the carbon footprint of bottle transport in particular.

– It should also be noted how the gradual generalization of differentiation by the GI, at least at the European level, would risk reducing its scope by bringing it closer to a standard, especially in view of the now fairly common reference to the "brand x grape variety" pair on the label.

Therefore, not being able to supply this type of circuit in sufficient volumes exposes it to being replaced by a competing offer, especially internationally. In such a context, a lack of wines is making it difficult to win back any markets not supplied in the previous year. This is all the more important since, in case of insufficient forecast availability, the "natural" lever to maintain its commercial capacity in volume is to increase the yield, for example, by cutting longer, or by playing on the management of the individual marketable volume.

However, this capacity seems to be shrinking as a result of the downward trend in wine-growing yields since the 2000s. This decline is particularly due to vine decline.

Thus, it is best not to allow a surplus to accumulate while simultaneously avoiding the risk of running out of products commercially, knowing, however, that wines that are marketed as part of the luxury economy generally escape this second constraint. Hence, the legitimate question on the ways and means to stay on this ridge line, knowing that training industry players and an adapted marketing are probably part of the panoply of levers to stay there.

But why preface this book with a focus on major wine-growing specificities? It seems to me that, through the articles it brings together, this book sheds a particularly interesting light on a question that could be summarized as follows: do the characteristics of the vine, those of wine and its market functioning justify an equally specific approach to training and marketing in a broad sense for this sector while it appears today at the crossroads between, on the one hand, general contextual vectors such as increasing global urbanization, the effects of climate change or the permanent evolutions of ICT (Information and Communication Technologies) and, on the other hand, disciplinary progress, for example, in neuroscience or in the social sciences in terms of interculturalism?

Patrick AIGRAIN
FranceAgriMer

Introduction

I.1. Wine management and marketing: a field of analysis and major questions

This second volume of *Wine Management and Marketing*, following the first volume published in 2021, proposes to present new studies on issues in the sector, on the competitiveness of companies, and on organizations in a competitive but highly regulated market context. Therefore, it approaches the wine industry in different dimensions, with plural approaches, authors specializing in different disciplines of the social sciences, economics and management, and different angles and objects of study, in order to try to shed light on its complexity. Thus, the subjects of the various chapters proposed concern organizations or companies, often specialized in the production or trade of wine; they may also concern the sector or industry as a whole. In all cases, the authors, who are researchers strongly committed to this sector, were then able to consider many industrial specificities that give the industry a special place within the agricultural and agri-food industries. The bias here is once again to multiply approaches beyond the frontiers of marketing sensu stricto, sometimes management sciences or economics, since they shed light on wine markets and the impact of certain actions on the general economy and the competitiveness of the sector.

Here, we will not discuss the specificities of the wine sector in detail, which is extensively presented and analyzed in the first volume in 2021. But it is clear that particular characteristics mentioned then, such as the regulation mode historically established in Europe and based on a single definition, the hypersegmentation of marketed products, and the late but intense globalization of markets over the last two

---

Introduction written by Hervé HANNIN, Carole MAUREL, Foued CHERIET and Paul AMADIEU.

decades naturally contribute to inducing marketing and management issues specific to the sector. We could add that the French marketing sector, in particular, has a belated relationship with this discipline, which was imported from the United States in the second half of the 20th century. In any case, it seems useful to us to continue to address, from an always interdisciplinary perspective and within the social sciences, some of the major questions that drive the sector and that constitute the authors' research subjects.

As with the first volume, it is again a question of offering the work of researchers in an accessible and easily understandable version to nonspecialist readers, interested in the French wine sector. Thus, the authors, who are mostly teachers as well as researchers, have tried to use structured modes of presentation, as well as a pedagogical style and approach, but it is by no means a question here of courses or didactic presentations, let alone recipes supposed to guarantee success in company management. While the coordinators and the majority of the authors are still members of the Montpellier research cluster, structured around the Institut Agro Montpellier[1], INRAE and the University of Montpellier, some colleagues from other institutions, such as Montpellier Business School or other centers, particularly in Bordeaux, have been associated. It was important to us that beyond delivering research from our Montpellier units, this publication also illustrates some collaborations among the multiple partnerships existing between French research centers. The quality of these relationships, which does not exclude some competition as well as has real emulation, is indeed fundamental. We are delighted by the richness of these structural or more ephemeral research networks, which make it possible to focus our energy on the major challenges that the sector faces.

Approaches and results of recent work are presented in chapters of about 20 pages, on various themes in their disciplines and objects, and converge towards themes related to strategies and competitiveness of companies and the sector, faced with classic challenges (market adaptation, management, etc.) or those which have appeared more recently (climate change, Big Data, etc.).

Then, there is a renewed ambition to provide tools, reading grids, illustrations and reference case studies to the various actors in the sector, and without refraining from using them for managerial and operational recommendations.

1 Montpellier SupAgro, created in 2007, became the Institut Agro Montpellier on January 1, 2021, one of three schools integrated into a single institution, the Institut Agro.

I.2. Why a second volume? Objectives, additions, new themes and new authors

There are many questions which are constantly renewed. The first volume had naturally been unable to answer all of the issues raised, as wide as the panorama presented at the time could have been. Some themes were not taken up, others were elaborated upon and some others were finally addressed. Thus, questions specific to the sector, ways of approaching climate change, the "coopetition" between organizations in the same region and the particular role of innovation in a sector, which has made tradition into the DNA of its institutions as much as its primary axis of communication, had been particularly analyzed in dedicated chapters.

Some themes, which were also discussed in 2021, are explored in greater depth, for instance, export strategies of cooperatives, consumption and markets open to export in Africa, or certain modes of communication, such as *storytelling*, and spaces closed or opened by the Évin law. Finally, we address new themes in this volume, with illustrations and appropriate application cases, which are in various fields of the social sciences: economics with environmental changes or the concept of "cluster" (with the Montpellier case), strategic marketing (the positioning common to certain associated terroirs), operational marketing (such as communication or wine tourism) or management, whether that of stocks or that of digital data.

In most cases, the reader will find food for thought to help them to better understand the industry's current issues, which are often the result of critical situations. Thus, the organic wine market, which had nevertheless experienced a historic take-off in the last decade and which is now encountering new difficulties, is approached here from the original angle of "willingness to pay", a valuable behavioral and experimental economy tool developed recently. Similarly, the indispensable but always difficult dialogue between research and professionals can be enriched by the approach provided by the "cluster" concept. How can we not make crises experienced today resonate against the background of product "deconsumption", which has not been a basic necessity for a long time, with the difficult but happy initiatives of building strong collective positioning and images as in Beaujolais or in the volcanic Loire region?

The issue of wine estate acquisition is of particular interest, not only because of the emblematic "Petrus" case of which is discussed in particular, but also because of questions raised by the evaluation of price and the "emotional" value of such companies.

I.3. Recent context of the wine sector and "new" challenges

These motivations for a second book were encouraged by a particular context, generating a series of commercial as well as organizational and even institutional challenges for sectoral actors. Among these contextual elements, several are directly related to the characteristics of the sector and the product.

The wine sector is deemed to be fragmented between many players of different sizes and governance. Indeed, listed groups coexist with smaller structures, including small family SMEs. All of these actors belong to different families, are at different stages of the value chain and have varied governance: farms, anchored in a family for generations or on the contrary composed of financial investors, cooperatives and traders. Organizations are legion in the sector, depending on professions, wine types or producing regions. There is a frequent temptation to merge, concentrate or simply organize to better weigh in the competitive universe and try to better adapt to it.

Thus, Chapter 10 presents the process of finding a common positioning to promote wines of four vineyards of the Massif Central. This implementation is fraught with difficulties. Indeed, a collective project implies uniting different entities that must act together and these entities themselves have, within them, actors with different and sometimes contradictory visions of the project. Bringing these actors together proved to be a difficult challenge, but one that was able to succeed by force of will and participatory method.

Chapter 12 proposes another example of collective construction, resulting from the collaboration between a regional brand, Vins d'Alsace, and a Bacchic society, the Confrérie Saint-Étienne, showing the benefits that can be derived from this collaboration.

And to go even further in this collective dimension, Chapter 11 shows how an old but still very active cluster, the Montpellier Vine and Wine cluster, by bringing together many actors in the sector, including teacher-researchers, students, experts, policies, collective bodies, producers, etc., and making them interact, makes it possible to increase the competitiveness of companies and the sector as a whole.

As another illustration of collective strategies, Chapter 16 presents the particular situation of cooperatives and cooperative unions facing the same export problem. It shows that, overall, cooperatives and their unions export less than other companies in the sector. This phenomenon is more pronounced for cooperatives that appear less well adapted than cooperative unions to export their wines. Exporting requires a great deal of adaptability.

In any case, in domestic or foreign markets, the need for a clear positioning remains a prerequisite for any collective communication or distribution strategy or policy to be implemented. This is made clear in Chapter 9, which explores the case of the launch of Beaujolais rosé to respond to new consumption habits and to cope with the reduction in red wine consumption.

Wine is a product frequently rooted in the terroir, the quality of which is difficult to assess without tasting it, and is often with a weak and difficult to discern differentiation. All of this forces producers to implement strategies to distinguish themselves from the competition and meet the expectations of French and foreign consumers. It also benefits from a strong social image (tradition, conviviality, luxury product for certain categories) and is subject to high regulation in France and abroad, whether in terms of production (appellation system), marketing (e.g. state monopolies in certain countries of the world) or communication. All of these characteristics encourage companies in the sector to be flexible and to develop general strategies, particularly in marketing, while integrating these sometimes restrictive specificities, as in the case of labeling with regard to regulations. It is then that different options, proposals and communication techniques can be implemented, in compliance with legal and specific rules like the Évin law, and also by adapting to consumers' expectations and cultures. Thus, Chapter 6 explores the use of mental imagery (or "imaging instruction") for the communication of wine businesses in compliance with regulations. When a company wishes to communicate internationally, it may face particular regulations as well as different cultures. In this context, Chapter 8 considers how to adapt communication according to the countries where it is carried out. A "semiotic translation" of advertisements is recommended according to the countries where they are broadcast. In a more traditional dimension, Chapter 5 shows how storytelling and the narration of authenticity can build an identity and legitimacy for new-winemakers. This allows them to create emotion-based connections with their customers, prospects and the wider environment.

The objective is still the ability to meet market expectations, a market of consumers who are often not very knowledgeable, yet who determine profits made by companies in the sector by their choices, behaviors and purchases. Thus, for many years, we have observed a set of changes, both in the domestic and global market. The decline in consumption on French soil makes it essential to be present on international markets, and exporters sometimes have to be bold to develop in "emerging" markets, as is the case in Africa, a market that presents interesting growth opportunities. To address it, wine companies must communicate in an intercultural approach, as shown in Chapter 13, which proposes a reflection on marketing orientations to be adopted towards African consumers. Conquering this market also requires the development of relationships of trust with local importers. Chapter 15 shows how good social interaction between actors can build this relationship of trust by adopting a transdisciplinary stance.

But, more generally, how can we ensure that the products put on the market can satisfy consumers? Chapter 7 presents a study that measures the value of organic wines based on consumer's willingness to pay. If they are willing to pay more for an organic wine for the same quality, producers' efforts to meet market expectations are truly rewarded.

It then appears possible to better understand consumers, and other tools could well contribute to this in the coming years. Thus, Chapter 2 presents opportunities offered by digital transformation and Big Data to improve marketing policy and customer relations. Indeed, these digital technologies have experienced a particular boom in recent years, and undoubtedly further strengthened with the health crisis. Still very little exploited, they present great development opportunities for companies in the sector generating real challenges and opportunities for actors in the sector (e-commerce development, reflection on the integration of Big Data into the strategy of companies including SMEs, etc.).

Around these actors is a dense institutional ecosystem involved in the regulation, financing and visibility of companies in the sector (inter-professional organizations, INAO, appellation unions, federation of wine and spirits exporters, FranceAgriMer, financial institutions offering dedicated centers, etc.) that influences the heart of the Business Model of these companies, offering them particular levers to adapt to market and societal demand. Thus, Chapter 1, focusing on the central issue of inventories from a survey of professionals in the sector, presents how inventories characterized by their large volume in viticulture can become a strategic and financial tool at the service of objectives pursued by a company.

Dedicated to another management issue, more heritage-related, that of the sometimes astronomical valuation of prestigious estates, Chapter 3 shows, through a concrete case, that viticulture attracts capital beyond the purely financial logic because of an emotional value accepted and even desired by investors. This phenomenon does not only affect the most prestigious estates. It also concerns those who are part of a long-term sustainable development strategy, by producing wines whose quality is praised by the market and by adopting practices recognized as virtuous. These companies are therefore not in a short-term financial logic. They have a lower financial constraint. They can then have capital at a lower cost than in other sectors, at an equivalent risk. The emotional value attributed to companies in the sector thus makes it easier for them to have the necessary financing for innovations and adaptations expected by the market and society. Well-invested, this capital can allow, for example, to engage in international development strategies, as explained in Chapter 4, which analyzes the international position of French viticulture and evokes proposals that would promote French leadership. It may be appropriate to build strong brands to compete on the international market using opportunities offered by new technologies and "protean" innovation.

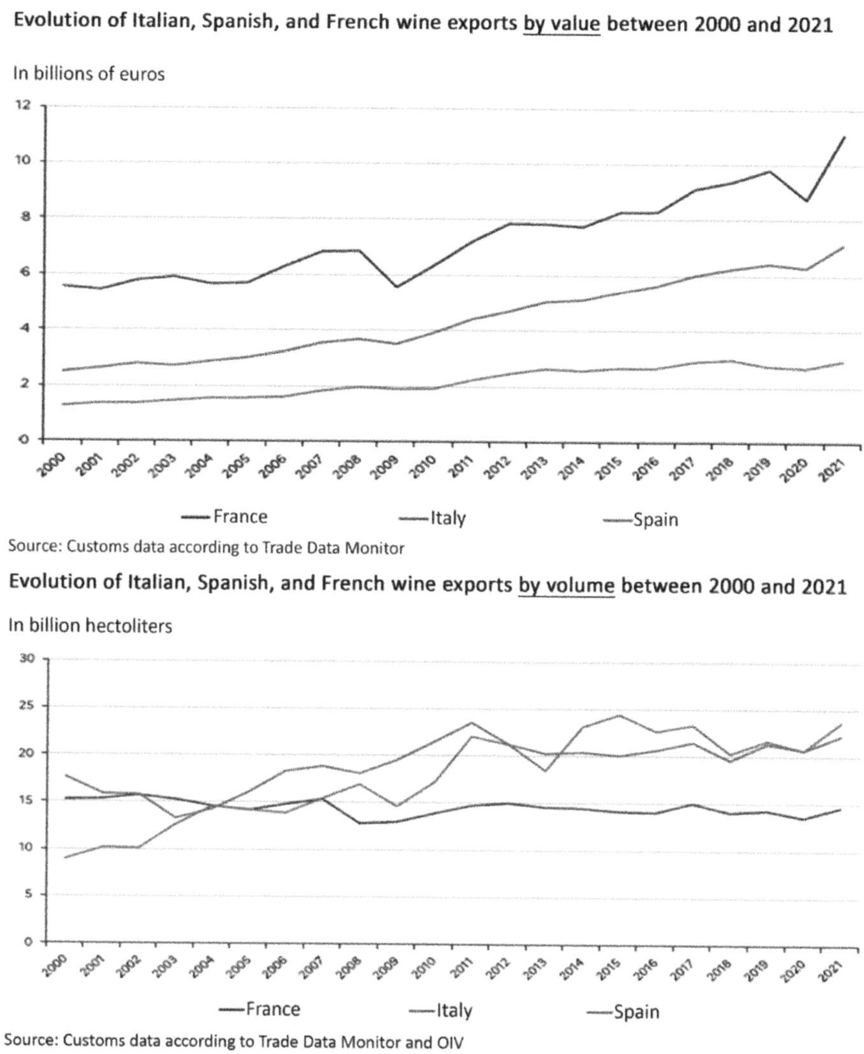

Figure I.1A. *Evolution of world wine exports by country and by production areas (source: data from FranceAgriMer 2021/Vins). For a color version of this figure, see www.iste.co.uk/cheriet/wine2.zip*

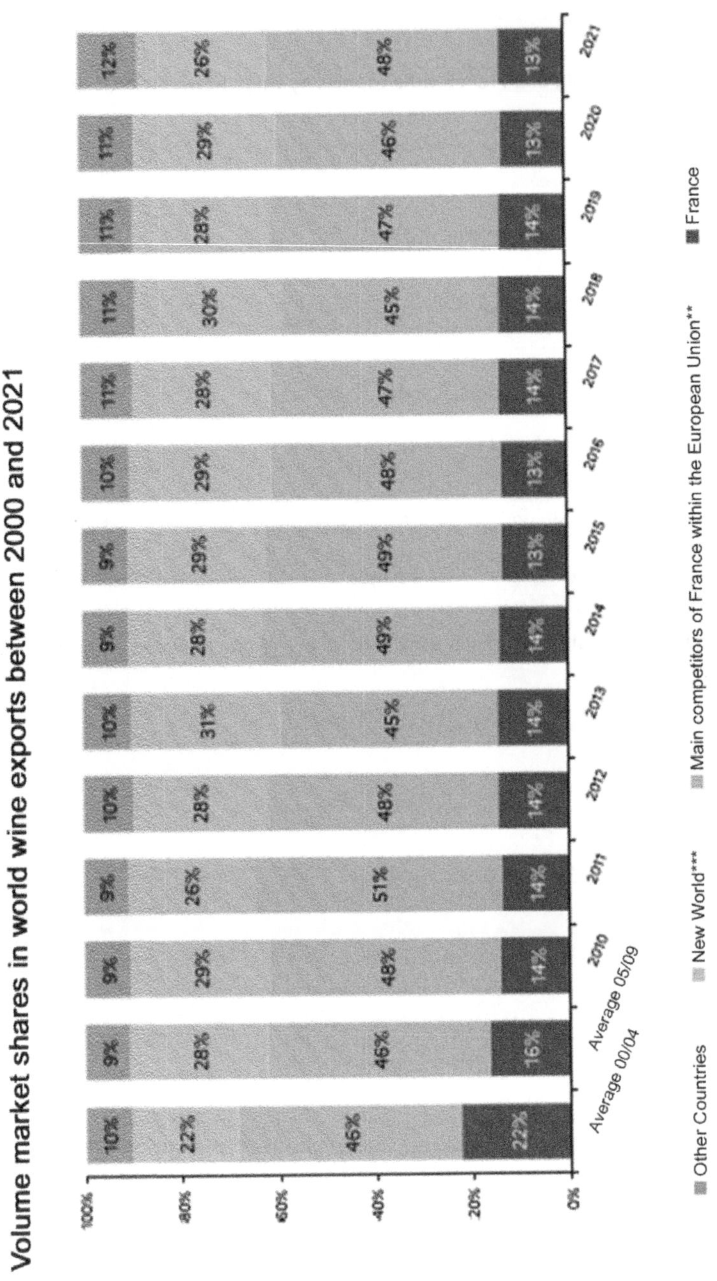

Figure I.1B. *Evolution of world wine exports by country and by production areas (source: data from FranceAgriMer 2021/Vins) (continued). For a color version of this figure, see www.iste.co.uk/cheriet/wine2.zip*

France has historically been one of the leaders, even if its place is regularly challenged by traditional producer countries Italy and Spain (see Figure I.1) as well as by other producer countries offering an offer of increasing quality from year to year. Consumer habits are changing (see Figure I.2), whether on the domestic market or on international markets, with more occasional consumption, a stronger experiential dimension, a decline in French consumption in volume, and increasingly demanding foreign consumers. Consumers' concerns and expectations are also evolving towards a "healthier diet", in line with environmental and societal issues, in addition to the more traditional issues of quality and price. Thus, part of the demand is waiting for a more environmentally friendly production, which has notably contributed to the rise of organic wine (see Chapter 7) following a curve that is now reaching limits by meeting new competition.

Figure I.2. *Changes in consumption habits (source: 2002 Sowine/ Dynata barometer data[2]). For a color version of this figure, see www.iste.co.uk/cheriet/wine2.zip*

Finally, the wine sector, like all other sectors of activity, must deal with changes and shocks affecting society as a whole, some of which affect companies in the wine sector more strongly, in particular climate issues, as well as Trump taxes and Brexit

2 See: https://sowine.com/blog/resultats-barometre-sowine-dynata-2022.

a few years ago, linked to two key destinations for French wines abroad. More globally and systemically, the world regularly faces economic and financial crises, and more recently the Covid-19 health crisis, which has shaken the whole world and which has required companies to be resilient, to adapt and even reinvent themselves, as shown in Chapter 14 on export professions. It presents how professionals in the sector, faced with a crisis, have been able to demonstrate their ability to adapt. Through a survey, Chapter 14 explores how export managers maintained their business during the Covid-19 health crisis and showed flexibility. This study also focuses on management in the post-Covid-19 phase, and in particular on how to sustain the tools developed to deal with lockdowns. This sustainability can then be achieved by combining existing skills to develop others.

I.4. Sixteen chapters on new strategic and marketing trends

Thus, in order to shed light on all of these issues and to allow companies and institutional investors to benefit from keys to reading the current context and the issues it generates, this book proposes a set of 16 chapters, grouped into four main thematic and disciplinary sections.

The first four chapters address themes that are relevant to strategies that echo the major cyclical trends that impact the sector. Thus, it will discuss the strategic role of wine inventories (Chapter 1), digital transformation and Big Data in SMEs in the sector (Chapter 2), and wine land valuation issues (Chapter 3) as well as the international strategies of wine companies in response to trends and changes in their international environment (Chapter 4).

The second part is devoted to the marketing component, through four chapters focusing on current trends related to the context in which the sector evolves: storytelling (Chapter 5) as a vector of differentiation, two chapters on wine communication and advertising with regard to the evolution of regulations (Chapters 6 and 8) and a fourth focusing on the "organic" attribute and its link with quality and perceived value (Chapter 7).

The third part focuses on collective strategies and terroir dimensions through four chapters. It will be a question of terroir and collective image (Chapter 9), collective positioning with the emergence of a collective brand (Chapter 10), synergy between a brand and a Bacchic society on the same wine region (Chapter 12), and the presentation of a cluster promoting synergies between sectoral actors and the academic world in another wine region (Chapter 11).

Finally, the last part discusses the international context and the challenges facing sectoral players. The first chapter presents the evolution of export trades following the health crisis (Chapter 14). Two chapters then focus on the role of French wines on the African market, with a focus on the welcome given by African consumers to Bordeaux wines (Chapter 13) and on the role of trust in the B2B relationship in this same market (Chapter 15). Finally, the last chapter discusses export strategies of wine cooperatives for export (Chapter 16).

PART 1

Strategies, Regulations and Finance

1

Inventories in the Wine Sector: A Strategic Question

1.1. Introduction

The wine sector is currently facing multiple challenges. In addition to Brexit and Trump taxes in France, for example, the Covid-19 crisis increased the inventory weight, even though it was necessary to free up storage space for the new harvest. Climatic and commercial hazards exacerbate tensions on wine inventories: scarcity of products accompanied by situations of excess inventories.

Wine inventories are a key issue in this sector (Koulayom and Chopov 2007). The operating cycle of a wine business is very long, sometimes several years. It results in holding a high stock, which implies mobilizing costly resources in terms of human and financial resources, infrastructure and logistics. Maturing can range from a few months for early blossoms to several years for wine kept for maturing (Goncharuk 2017).

Inventory implications cover different dimensions: strategic, accounting and financial. They raise various kinds of questions: valuation of inventories for high-end wines and champagnes, logistical and cost issues for small producers and wine merchants, and valuation of assets for mergers and acquisitions. Several situations must be considered. According to a study carried out by Crédit Agricole, wine inventories accounted for 273 days of sales in 2019 (256 in 2016), compared to 77 days in the agri-food sector as a whole.

Chapter written by Carole MAUREL, Françoise PIERROT and Foued CHERIET.

The wine industry is experiencing low inventory turnover. Under these circumstances that are directly linked to the nature of the product and its production process, wine companies must face financial implications of such a large but relatively "inert" asset. The Crédit Agricole study highlights that equity represents 37% of total assets (43% for the entire agri-food industry) and that ROCE[1] is structurally lower due to large inventories, in line with an important role played by inventories, as explained by Estrade and Koulayom (2009). The inventory level influences costs, and therefore leverage and profitability, and the ability to cope with changing demand.

The wine inventory issue is also contingent on vineyard size, appellation specifications, yield and climatic impact such as drought, hail and frost, weakening the performance and durability of these companies. In fact, wine inventories can be adjustment variables to cope with the impact of climate change (Bukenya and Labys 2007). Hail, frost and other climatic hazards can have a considerable impact on harvests, and wine inventories are a way to regulate the volume of wine available for sale. In total, wine companies must specifically manage inventories, in order to find the best compromise between production process and environment-related constraints, to reach expected profit levels, while being flexible to demand expectations, and sustain the business. Inventory management implies that companies face a "trade-off between limiting risk by having a higher level of inventory and limiting inventory cost" (Michalski 2008): wine companies must find the right level of inventory to hold to meet demand and ensure that vats are empty and cellars available for new harvests. Inventory operating costs are related to storage, insurance, transportation, obsolescence, waste as well as inventory deterioration (Scherr 1989).

To our knowledge, inventory management analysis is widely applied to industrial sectors, but no previous study focuses on inventories in the wine sector or presents inventory policies, company behavior and inventory performance in this specific industry. Bukenya and Labys (2007) focus on the impact of wine inventory fluctuations on wine prices, while Estrade and Koulayom (2009) and Koulayom and Chopov (2007) provide an overview of the accounting and financial problems of wine companies, including inventories. Finally, Aytac et al. (2016) study the impact of working capital and the financial valuation of inventories on the profitability of companies.

The current context shows that inventories play a crucial role in the longevity of wine companies. But how are inventories of wine companies determined? What strategies do they adopt? What are the consequences of management of wine

1 Return on capital employed: return on invested capital, measured by the ratio "operating income on the sum of shareholders' equity and medium- and long-term debts".

inventories? To answer these questions, we conducted an exploratory study focusing on still wines, in order to shed light on wine inventory determinants, storage behaviors and strategies of wine companies, in relation to their performance. This study combines interviews with wine business executives and experts related to the wine sector (accountants, consultants, researchers, etc.).

In this contribution, we seek to identify different strategies implemented in the wine sector, in order to analyze their specificities and implications. The first part presents "wine" products and wine sector specificities, justifying the need to carefully study wine inventories. We also discuss inventory determinants and their relationship with several dimensions of a company. The second part presents our empirical approach and the main results obtained. We will describe wine company behavior in terms of wine inventory management as well as the main issues they face.

1.2. Inventories in the wine sector: definition, determinants and implications

1.2.1. *Inventories and wines*

Wild (2017) described inventories as "the fluid that lubricates supply chain operation, as they separate demand from supply". Barnes (2018) defined it as "the accumulation of physical goods as they pass through a process or network of operations". They are generally classified into three categories, raw materials, work in progress and finished products, the level of which depends on predictability, fluctuations in demand, unreliability of supply, price protection, quantity discounts and lower order costs (Muller 2019). In accounting, they correspond to current assets held in order to be sold or used in the production process during the operating cycle, the current duration of which is equal to 12 months in most industrial sectors. However, wine production takes longer, and much longer in some cases.

Beyond the commercial challenges of managing finished products and vintages, the issue of stocks concerns work in the vineyard (work in progress inventory and the concept of advances on crops), winemaking, and the aging and maturation of wine. Each stage has different durations and entails financial, strategic and commercial risks of a different nature and scope.

Winemaking involves different steps of processing, from the grape to the bottle: once the ripe grapes are harvested, they are destemmed and crushed to give a grape juice called must. Then, the must is fermented with yeast. The next step is clarification to remove solids by gluing or filtering. The last step is to age and bottle the wine before selling it. Each step results in different inventory types.

The wine production process, carried out by winegrowers, cooperatives and wine merchants, is reflected in accounting by several categories: (1) wine, bag in box, (2) dry matter (bottles, stoppers, cartons), (3) bulk wine and (4) other supplies. Our study focuses on the first and third categories. Wine inventory management thus consists of finding the right rotation of wine inventories from vats to barrels and to cellars, where bottles are stored. This rotation depends on the level of sales and a company's ability to sell wine from previous harvests, keeping in mind the need to have empty vats before the new harvest, in order to be able to welcome the new vintage.

1.2.2. *Determinants of wine inventories*

According to Qian (2018), inventory management literature studies either optimal inventory policies or the behavior of companies in managing their inventories, and links between performance and inventory. Studies specifically devoted to inventory determinants are rare. Several models have been developed to minimize all costs related to inventories and their financing. Lieberman et al. (1999) present several theoretical frameworks to explain what factors determine the level of inventories held by companies. Classic inventory theory aims to define the optimal level through economic order quantity (EOQ) (Choi et al. 2018). They showed that the level of inventories depends on product and process characteristics such as demand, installation costs, product costs and cost of storage time.

Choi et al. (2018) also demonstrated that several costs are related to inventories, namely purchasing costs, ordering costs, carrying and handling costs, out-of-stock costs and quality cost. This theory also aims to determine the level of inventory security. This level is positively related to the variability of demand and product quality, the desired level of service and the time required to replenish supplies (Lieberman et al. 1999). On the other hand, it is negatively related to the cost of holding and future valuation of this security stock. The authors explained that the development of the just-in-time approach impacts the level of inventories with managerial decisions and control. Mathuva (2013) identified determinants at the company level such as time, sector and economy of inventories held. But to what extent do these determinants make sense in the wine sector?

1.2.2.1. *Company-related determinants*

Studying wine inventories involves considering the diversity of companies involved in the production and marketing process. In France, for example, the wine sector is globally composed of four groups of companies: independent winegrowers, "pure" traders, wine merchants and cooperative wineries. In parallel with the concentration of the sector, many traders are buying vineyards to establish their sales entity and control quality, blurring the boundaries between these different families.

Therefore, most wine companies are affected by different inventory management types, depending on their role in the value chain and supply chain strategy (Garcia et al. 2012; Moccia 2013). The literature does not focus on factors impacting inventory management. However, stakeholders in the wine sector consider inventory management, depending on the situation in a different way: bulk or bottled wine production, viticulture, wine trading, etc.

1.2.2.2. *Sector and economy-related determinants*

The way in which wine companies manage their inventories depends on constraints related to research, the quality and time required for the wine to be ready for sale, as well as on institutional factors. Recent years have shown that such events can impact inventory management and levels: political and commercial environment (customs duties, Brexit, etc.), as well as health (Covid-19). More recently, events in Ukraine have contributed to higher dry matter and logistics costs.

Wine quality is a key success factor. In many wine regions, companies are fighting to improve wine quality (Bukenya and Labys 2007). For 40 years, producers in Languedoc (France) have moved from a volume strategy to a quality strategy to cope with internationalization and increased competition. Blake et al. (1998) explained that improving quality leads to adapting accounting rules: valuation of quality grapes; inventory management between fermenting and bottling; and tax implications. However, even if a few years of aging seem better for quality, Dimson et al. (2015) demonstrate that the best financial return of wine comes from a specific combination of the number of years of aging and the level of quality of the vintage considered, namely "young aging – quality vintage".

Inventories are also impacted by other sector-specific factors, such as the institutional context. Indeed, another determinant is based on the existence of *Appellation d'Origine Contrôlée* (AOC), which imply legal specifications binding the period from which the wine can be marketed. Some specifications may prohibit companies from selling their wine before a given date. This implies an additional need for short-term resources to cover this storage period. For example, the AOC Pic Saint-Loup requires red wines to be aged until July 31st and marketed no earlier than September 1st of the year following that of harvest.

The AOC system also allows wine companies to constitute an individual complementary volume. In periods of good harvests, they can store 20% (instead of 10% until then) of their harvest each year for up to half a year of harvest over three years. This system was developed to limit the effect of climate change, which can seriously compromise the sustainability of wine businesses. This seems to be an interesting and economical form of insurance against climatic hazards, but it involves having the buildings and vats necessary to store this surplus wine, and monitor these wine inventories (quality).

The French system also includes the principle of "en primeur" wines. It is similar to futures, derivatives that allow you to buy wines before they are bottled and at a cheaper price, two to three years before the wine is released. This system started in Bordeaux and has spread over time to other producing regions, because it allows wine companies to secure part of their turnover. Even if this means that winemakers take care of wine storage for a few weeks, they can quickly benefit from a direct cash flow intended to finance working capital, with price being partly paid in advance and the balance (VAT, delivery costs) on delivery. Selling "en primeur" wine allows for reducing stock weight for companies, from a financial point of view as well as in relation to the necessary infrastructure and logistical challenges.

1.2.3. *Consequences of wine inventory management*

1.2.3.1. *Wine inventories and performance*

Wine inventory management faces various limitations that must be countered to optimize the level and value, because they in turn influence a company's performance. In this case, performance can be defined as a set of financial and nonfinancial indicators reflecting the achievement of objectives and results (Lebans and Euske 2006).

Numerous studies have studied the link between companies' performance and their inventories. Koumanakos (2008) showed that the higher the level of inventories, the lower the rate of return. On the other hand, the literature agrees that a more aggressive working capital policy has a negative impact on profitability (Aminu 2012). Studies on this issue show mixed results (Mathuva 2013). Capkun et al. (2009) found a positive correlation between inventory performance (average annual inventory/sales, turnover indicator) and, on the one hand, financial performance and, on the other hand, the valuation of US manufacturing companies.

The wine sector is heavily limited by inventories due to the production process that goes beyond a classic operating cycle chronology. These inventories become a de facto financial determinant of performance (company valuation), in addition to a production characteristic or a commercial tool. According to Esteban-Rodríguez and Climent-López (2018), inventory plays a different role in profitability: sales predictability is easier in the generic market than in the dedicated product market. In these dedicated product markets where variations are more uncertain, companies fearing lack of sales tend to reduce inventory levels, whereas, in generic markets, we can afford to have more inventories.

In the particular case of wine, profitability is also impacted by inventories of long-keeping wines. As aging generally improves wine quality, aging wine inventories are also part of the company's value and performance. Aytac et al.

(2016) studied the impact of working capital requirement management on corporate profitability and found that the cash conversion cycle, measuring the time it will take a company to convert its stock into cash, does not significantly impact profitability, whereas theory anticipates a negative impact. Aktas et al. (2015) explained that due to different costs and benefits, the relationship between working capital requirement and business performance is not linear.

Ideally, accounts payable should offset accounts receivable and inventories should be financed by short-term debts or a stable surplus of resources. The structural part of the working capital must be distinguished from its temporary part. The inventory financing issue appears even more crucial in SMEs, so that banks have developed specific solutions for the wine sector in certain producing regions. They require inventory to be used as collateral against a loan, implying another inventory valuation problem (Blake et al. 1998).

1.2.3.2. *Wine inventories and company strategy*

Implementing the strategy requires specific inventory management (Kotler 2002). It is a question of "balancing the conflicting economy of not wanting to hold too much inventory" (Adeyemi and Salami 2010), in other words a trade-off between limiting the "out-of-stock" risk and limiting storage cost (Michalski 2008). A company's value seems to depend on this trade-off through the impact on working capital and therefore on free cash flow (Michalski 2008).

As explained by Bukenya and Labys (2007), among various existing storage models, one key is to determine the optimal level of inventories for which marginal revenues become equal to the marginal cost of holding inventories. Nevertheless, two difficulties emerge in the wine sector: assessing storage costs even though farming is an integral part of the production process, and assessing the "value" of inventories, while some companies aim to keep wine to increase its value.

We may also wonder whether the level of inventories, due to the nature of inventories, activity and context, impacts the strategy, where inventories are associated with a certain level of uncertainty due to volume variations which are themselves dependent on climate change that is intrinsic to agricultural activities. Finally, inventory level and value influence a company's choices in terms of distribution channels, production trajectories and investments that are linked to these choices.

1.2.3.3. *Wine inventories, financial statements and valuation*

In financial statements, inventories appear as a current asset in the balance sheet. Accounting raises the question of the value of inventory and, therefore, the accounting standard used to measure it. Several standards exist, as each country develops its own, with the exception of those that use international standards. Generally, inventories are valued according to acquisition cost (raw materials and goods) or production cost (work in progress and finished products) depending on the asset in question. As this value is only adjusted in the event of a potential decrease, the increase in value resulting from wine aging is not recorded in accounting books.

Koulayom and Chopov (2007) highlighted the difficulty for accounting standards to represent a true picture of the situation when wine has to age for many years. To explain this, they cited Rodriguez and Ocejo (2002) who considered that the cost of initial recognition had to be adjusted. This is justified since the quality of wine increases with aging, but this aging leads to other accounting problems. Inventories are classified as current assets intended to be disposed of as part of the operating cycle. Aging can last several years, and the accounting image no longer reflects the reality of this aging, because accounting standards are not adapted to this wine-growing specificity. For example, some Australian companies choose to classify aging wine inventories as noncurrent assets (Juchau 1996), in order to provide users with a more accurate and realistic accounting representation of the value of their wines.

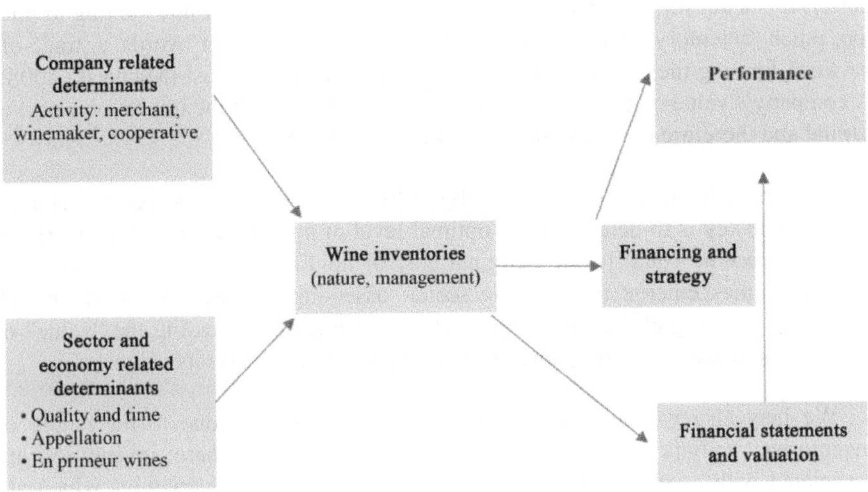

Figure 1.1. *Details of determinants and consequences of wine inventories based on results obtained*

This literature review aimed to explore the nature, determinants and consequences of inventory management in wineries. It allowed us to highlight some specificities of wine inventories. Indeed, we explain that inventories and their management depend on the nature of the wine and limitations related to the production and distribution process, as well as the role of a company in the value chain. Finally, the level and value of inventories throughout the year impact a company's performance as well as its value and the strategy that will be implemented by the company in terms of the production and distribution of its wine. Figure 1.1 presents a summary.

1.3. Exploratory study on inventory management in the wine sector

1.3.1. *Methodology and data*

This exploratory qualitative study is based on data collected during interviews with various stakeholders in the wine sector. We conducted 16 semi-structured interviews either by phone or face-to-face, in France and the United States (California) between October and November 2018, with several parties:

– five managers of French wine companies from different production regions (telephone): three in Languedoc, one in the Rhône Valley and one in Burgundy;

– three California winery managers (face to face);

– two employees of French banks specializing in the wine sector (face to face) in the Rhône Valley;

– two French accountants (face to face), including one from Languedoc and one from the Rhône Valley;

– one employee of a Languedoc association helping farmers to set up their business (telephone);

– two Californian researchers and one Spanish researcher (face to face).

Interviews lasted between 30 and 90 minutes each, or 865 minutes (about 14 hours) of cumulative recording. They were organized around different themes: inventories and their determinants, links between inventory management strategies and performance dimensions (accounting, financial and economic), and the role of inventories and tax issues. These interviews were transcribed and analyzed manually to preserve the quality of the information.

French respondents helped us get illustrations of inventory issues. The views of foreign interviewees, whether winemakers or researchers, showed us another conception of the role played by inventories in other renowned wine regions.

The interviews aimed to compare the approach of different actors, to identify key issues related to wine inventory practices in terms of valuation, financing and management. The interviews conducted were organized in four phases, as described in Table 1.1.

Stages	Contents
1	– Presentation of the company: size, activity (winemaker, merchant, cooperative, export activity). – Types of wines and consumers (supermarkets, CHRs (Cafés Hotels and Restaurants), wine merchants, etc.). – Group structure (parent company, subsidiary, etc.) or independent company. – Family business?
2	– Nature of inventories (bulk?, bottle?). – Weight in company assets. – Who holds the inventories?
3	How are inventories managed (optimal level, cost, insurance, etc.)?
4	How are inventories financed?
5	Problems and questions related to inventories (fiscal, financial, accounting, technique, etc.) identified by respondents.

Table 1.1. *Interviews, a five-step approach*

1.3.2. *Description of interviews*

	Regions	Size	Organization	Governance	Company start date	Export rate
1	Languedoc	20 ha	1 vineyard	Independent	1992	40%
2	Languedoc	135 ha	2 vineyards	Independent and family	1848	5%
3	Rhône	150 ha	1 vineyard	Independent and family	1929	50%
4	Languedoc	2,000 ha	5 vineyards	Subsidiary of an international group	1883	50%
5	Burgundy	25 ha	1 vineyard	Independent and family	2009	80%
6	Napa Valley	633 ha	2 vineyards	Subsidiary of an international group	1982	–
7	Napa Valley	20,000 ha	40 vineyards (international)	International family group	1982	–
8	Napa Valley	Unknown (< 10 ha)	1 vineyard	Independent and family-run micro-winery	1985	–

Table 1.2. *Description of companies surveyed*

Tables 1.2 and 1.3 present the main characteristics of the companies surveyed and others interviewed as part of our exploratory study. Table 1.2 focuses on wine companies, while Table 1.3 presents other actors.

	Regions	Role of the respondent
9	Rhône	Banker or bank employee
10	Languedoc	Banker
11	Languedoc	Chartered accountant
12	Rhône	Accountant
13	Languedoc	Employee of association helping farmers set up their business
14	California	Academic researcher
15	California	Academic researcher
16	Spain	Academic researcher

Table 1.3. *Description of other respondents*

In the first sample (Table 1.2), different company profiles are represented, from a large international group to a micro-winery, through the subsidiary of a listed group. Logically, French wine companies are, for the most part, older than Californian companies. In addition, we have also interviewed partners of wine companies to have a more global approach. We have not included Champagne as its production specificities encourage us to consider it separately. This chapter focuses on still wine. For the second sample (Table 1.3), the diversity of the respondents' origins aims to assess whether the strategies followed by French wine companies are also taken up in other wine-growing countries.

1.4. Results

We first present the behaviors and practices of wine professions and supplement them with information collected from other stakeholders.

1.4.1. *Determinants of wine inventories*

Wine inventory determinants are identified and classified into three categories: determinants related to the winemaking process, company characteristics and external factors. These elements demonstrate the specificity of wine inventories, which partly explain the difficulty in managing and achieving the necessary compromise and trade-offs, or in identifying optimal levels:

– Wine production process: inventories are directly linked to the wine production process as well as to the annual cyclicality of the activity (inventories are lower before harvest and higher after), to the constraints in terms of aging, bottling, and distribution and the nature of inventories (bulk/bottles in particular).

– Company characteristics: inventories also depend on the nature of the activity: bulk or bottled wines; who manages wine aging?; etc. Some companies sell both bulk and bottled wines. They must optimize trade-offs on the volume of bulk wine to be sold or bottled according to market prices, demand and cash requirements. Others produce only bulk wines and instead sell their gross production to merchants who grow, bottle and sell wine. According to answers during our interviews, inventories also depend on a company's financial situation: companies with a low level of profit and cash flow are less likely to hold inventories and need to sell wine faster. Inventories seem more problematic for small businesses where land is undervalued (lower loan guarantee amount), and for companies specializing in aged wines (higher working capital requirement). This raises the question of the extent to which the business model is impacted by a company's financial situation and size (impact on access to financial resources).

– External factors: in the event of climatic hazards, harvest is weaker and inventories decrease. Therefore, the volume of past harvests is also a determinant of inventories. More recently, the Covid-19 crisis has slowed down wine sales in domestic and foreign markets, and has generated abnormal inventory levels. This has led the French government to put in place exceptional measures such as a crisis distillation authorization, in line with the EU Delegated Regulation published in 2020[2]. Rules laid down by institutions with regard to alcohol, whether the government or wine institutions, also constrain the level of inventories.

The interviews highlighted some differences between French and US wine companies, where bulk wine is presented as a way to adjust inventories at the same time as the need for cash. In addition, long-keeping wine does not suit the taste of most US consumers, and wineries therefore do not have to bear the financial burden of long-keeping wine inventories. As an illustration, *verbatim* extracts are presented in the appendix (see section 1.7).

2 Commission Delegated Regulation (EU) 2020/592 of April 30, 2020, on temporary exceptional measures derogating from certain provisions of Regulation (EU) No. 1308/2013 of the European Parliament and of the Council in order to remedy market disturbances in the fruit and vegetables and wine sectors caused by the Covid-19 pandemic and the measures put in place in this regard.

Production process elements	Characteristics
Storage duration	It depends on the nature of the wine: longer for red wines and champagnes than for rosés and whites.
Storage type	Barrel, bottle, bulk wine arbitration: while barrel stocks are more or less always the same, bottle stocks are more variable and pose more problems.
Storage cyclicity	– Storing wine to be aged (emphasis on quality and premium wine) means having enough space and financial resources to support long storage → influence on product strategy? – Emptying vats to accommodate the new harvest and fill them a few months after the harvest.
Harvested volumes	Natural determinant influencing both the volume and quality of grapes, and impacting the entire winemaking process → winemaker impacted first, but indirectly the other actors in the value chain.

Table 1.4. *Wine production process and inventories*

1.4.2. *Wine inventory management and strategic and financial effects*

Our results also highlight the relationship between inventory management and other variables related to corporate financing, valuation, strategy and performance:

– Financing wine inventories: according to the interviews conducted, inventories appear to be financed by equity and/or short-term financial debts (overdraft, commercial paper). Bank employees indicated that the level of indebtedness depended on the wine production cycle. Banks value inventories on the basis of market price rather than on the basis of book value. Security is based on land (price and availability), the value of which depends on the region. Inventories themselves may be a guarantee against certain financing solutions. This raises the question of inventory valuation. Financing problems seem to strongly depend on the size and reputation of wineries: the constraint is higher for small and independent wineries. Interviews confirmed the literature.

– Wine inventory valuation: wine inventory valuation at production cost in financial statements does not seem to be an issue for respondents, who considered that fair value would be less favorable in terms of cash flow and tax implications. One of the respondents raised a problem related to the behavior of some large traders in high-value producing regions, in terms of valuation and overvaluation of their inventories, which will impact the entire sector, because their customers will have to pass this increase on to their price, or reduce their margin.

– Wine inventory management and strategies: the determinants of wine inventories and their management (costs, financing and degree of "liquidity") can also impact companies' business strategies, limited by their sales and storage capacity, on the one hand, and by their ability to finance their development strategies, on the other. While inventory management is determined by structural characteristics, the degree of inventory liquidity also depends on the financial needs and business choices of wine companies.

– Valuation, strategies and performance of wine inventories: inventories, an integral part of the working capital requirement, constitute one of the components of the value of a company, and inventory management strategies have effects on the commercial and financial performance of a company. This increases flexibility in short-term financial management (cash and valuation-accounting) and the use of inventories as an adjustment variable in performance measures, depending on objectives or stakeholders.

Figure 1.2 presents our main results by repeating the structure of determinants and consequences of wine inventories presented in Figure 1.1. As a result of our empirical study, we show that three, and not two, types of inventory management determinants can be considered. In addition, the nature of inventories appears more so as a determinant of inventory management than as the consequence of other determinants.

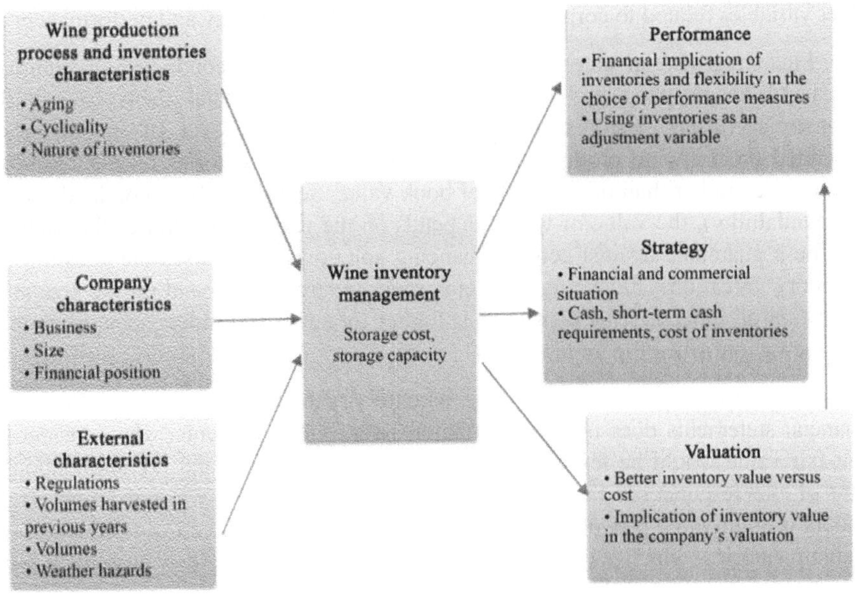

Figure 1.2. *Determinants and consequences of wine inventories*

1.5. Discussion

Wine inventories seem to respond to different managerial logics: operational (costs, storage capacity), financial (cash management and valuation) and strategic (choice of performance measures or commercial options). Certain characteristics of the product, the production process, and marketing reinforce the significance of wine inventory issues. Consequently, inventory management appears to be specific to the wine sector (Dimson et al. 2015).

Our results highlight the complexity of confirming whether wine inventories are a determinant of asset valuation, strategy and business performance. Nevertheless, the results observed in French companies show the role of inventories as a significant variable both in terms of cost and cash management, and as a determinant of business strategy (Mathuva 2013; Choi et al. 2018). Financially speaking, our results support an important role of inventory management as a particular and active variable in the valuation of many companies through financial and accounting statements (Estrade and Koulayom 2009).

From an academic point of view, inventory valuation appears more complex in the specific case of the wine sector (Koulayom and Chopov 2007), and requires adjustments in terms of measures as well as in relation to observed effects (Rodriguez and Ocejo 2002). This challenge is accentuated by the diversity of commercial strategies and the appellations and practices of firms: maintenance of the scarcity of supply, management of quantities and prices, nature of demand and competitors, and role of intermediaries.

The difficulty of assessing the strategic role of wine inventories makes it impossible to identify linear relationships between inventory management, strategy and performance (Koumanakos 2008), even if the value of this asset clearly contributes to financial valuations (Aminu 2012), with distinct effects depending on the nature of the wine stored (Esteban-Rodríguez and Climent-López 2018).

From a managerial point of view, our results show that, despite the role played by inventories in the wine sector, there is no specific treatment of inventories either in terms of specific accounting rules or in terms of the valuation of companies by banks. The particularities observed raise questions about the content of the accounting information relating to wine inventories as well as how to analyze it from a financial point of view.

In addition, the current context of crisis and risks related to climate hazards reinforces the role of inventories as an adjustment variable for companies as well as for public authorities and regional wine institutions. Constraints induced by the health crisis in terms of sales highlight storage issues (cost and capacity) and the need for an inventory management policy at the regional or even national level.

Interviews with US researchers suggest that inventory management is more oriented towards cash management in the United States, while results are more nuanced in France, with marked differences across regions. These two paths (comparison between French regions and international comparison) constitute interesting prospects for this research.

1.6. Conclusion

The objective of this chapter was to explore inventory determinants and strategies of companies in the wine sector. We conducted an exploratory qualitative empirical study based on a series of interviews with wine companies in France and the United States as well as other stakeholders (banks, support organizations, researchers, etc.).

Our preliminary results suggest that inventory management has certain specificities related to the product and the sector. Three groups of determinants influence practices: determinants related to the winemaking process, the characteristics of the company, as well as external factors. The implications of different strategies lead to differentiated effects in terms of financial valuation and solutions for financing wine inventories.

Our work suggests two main managerial implications: firstly, it highlights the fact that wine inventories are not a mere circulating asset, but rather a strategic and financial tool for wine companies. Then, our results suggest a macroeconomic treatment of wine inventories. They lead to the conclusion that further study of wine inventory management could provide interesting insights to help public decision-making in this area. Better knowledge for an active and targeted policy to go beyond a simple management of wine volumes: what level of public aid? For what type of companies? With which type of mechanism?

There are several limitations to this study. Combined with the low number of interviews, the concentration on two producing regions creates a bias in the scope of our preliminary results. Other producing regions such as Bordeaux and Champagne are experimenting with specific issues in terms of inventory management that could shed light on the empirical side of this study, related to the value of wine, the role of appellations and processional organizations, and the profile of companies. Similarly, we have not investigated certain cases of "active" wine inventory management: scarcity creation, speculation, arbitrations between vintages. Also, our study did not address the issues of heritage valuation of wine inventories, nor those of "logistics" management and trade-offs between the formats and outlets of stored products. For certain regions and appellations (Burgundy, Champagne, Bordeaux, etc.), this could have many specificities, thus opening up many prospects for future research. Finally, the question of intervention of public authorities in the regulation of wine inventories was not raised. The latest crises in the sector in France had shown that this is an important lever that should probably be better analyzed from a meso-economic and public action perspective.

1.7. Appendix

Determinants		Verbatim[(1)]
Wine production process and inventory characteristics	Aging	"We only have a small portion of our bottles that are stored for aging. We only have 10,000 10 year old bottles" (1). "We only have a few wines kept for aging. As I speak to you, we only keep the premium 2013 vintages for aging" (2). "Right now our marketing cycle is around a year" (4).
	Cyclicality	"The weight of each category of stock (bottles, barrels, bulk, etc.) depends on the season" (2). "The weight of each category of stock varies throughout the year and according to aging, bottling and distribution constraints" (3). "The stocks in barrels in the cellar remain constant, except for a few days when we go from one harvest to another [...], post-harvest, the vats and barrels are full" (2).
	Nature of inventories	"Bottles are the category of stock that causes us the most problems" (2). "Red wine is sold after 18 to 36 months of aging, we have on average two years of production in our cellars. As for rosé and white wines, the wine is sold after 6 to 18 months after the harvest" (3). "Bulk wine is the heaviest category of stock, financially speaking, but it does not cause us any problems" (4). "We have about 1.5 vintages in our cellars" (5). "White wine is stored from one year to 16 months, red wine is stored 19 months in barrels, and 6 to 18 months in bottles. Bulk wine is more flexible if we need cash, it constitutes a way to adjust inventories" (6). "Bulk is a way to trade-off between cash and storage" (8).

Company characteristics	**Business**	"We have two categories of stocks: stocks linked to the vineyard and stocks relating to our trading activity" (5).
	Size	"Inventories are a problem specifically for small wineries, certainly like in France, not for large ones" (8).
	Financial position	"It's difficult for us to keep wine to age because we need cash flow" (5).
External characteristics	**Regulations**	"In our appellation, we have to wait until July to start marketing the red wine that we harvested the previous year, contrary to what the Consumer Code says, namely December 15 of the harvest year" (2).
	Volumes harvested in the past	"The last two harvests have not been abundant, so we are not short of free space in our aging cellars" (3). "Inventories are a real issue. There is a problem of forecast depending on the harvest and market condition" (8).
	Weather hazards	"We need to be able to store for three reasons: wine for aging is valued at a higher price, it allows us to have tangible assets to obtain more financial resources, and it helps us deal with the consequences of climatic hazards" (5).

(1) The verbatim statements of all respondents except 6, 7, 8, 14 and 15 were originally in French. We have chosen to translate the original language in this table.

Table 1.5. *Identification of determinants in interviews*

1.8. References

Adeyemi, S.L. and Salami, A.O. (2010). Inventories management: A tool of optimizing resources in a manufacturing industry a case study of Coca-Cola Bottling Company, Ilorin plant. *Journal of Social Sciences*, 23(2), 135–142.

Aktas, N., Croci, E., Petmezas, D. (2015). Is working capital management value-enhancing? Evidence from firm performance and investments. *Journal of Corporate Finance*, 30, 98–113.

Aminu, Y. (2012). Determinants of inventories managements as a component of working capital in ensuring corporate profitability-A conceptual approach. *Research Journal of Finance and Accounting*, 3(11), 58–61.

Aytac, B., Hong van Hoang, T., Lahiani, A., Michel, L. (2016). Working capital management and profitability of wine firms in France: An empirical analysis. AAWE Working Paper, 201.

Barnes, D. (2018). *Operations Management*. Palgrave Macmillan, London.

Blake, J., Amat, O., Dowds, J. (1998). The drive for quality: The impact on accounting in the wine industry. *Journal of Wine Research*, 9(2), 75–85.

Bukenya, J. and Labys, W.C. (2007). Do fluctuations in wine stocks affect wine prices? AAWE Working Paper, 8.

Capkun, V., Hameri, A.P., Weiss, L.A. (2009). On the relationship between inventories and financial performance in manufacturing companies. *International Journal of Operations & Production Management*, 29(8), 789–806.

Choi, Y., Li, J., Wu, D. (2018). Applications of stochastic simulations in inventories management optimization: Cost accounting perspectives. *Journal of Applied Business & Economics*, 20(2), 184–204.

Dimson, E., Rousseau, P.L., Spaenjers, C. (2015). The price of wine. *Journal of Financial Economics*, 118(2), 431–449.

Esteban-Rodríguez, S. and Climent-López, E. (2018). Les appellations d'origine protégée du vin en Espagne, d'après la théorie des mondes de production : une approche quantitative. *Annales de géographie*, 2018/3(721), 279–306.

Estrade, Y. and Koulayom, H. (2009). Présentation de la journée Luca Pacioli sur les pratiques comptables et financières : vigne, vin, finance et comptabilité-un sacrilège ? *La Revue du Financier*, 176.

Garcia, F.A., Marchetta, M.G., Camargo, M., Morel, L., Forradellas, R.Q. (2012). A framework for measuring logistics performance in the wine industry. *International Journal of Production Economics*, 135(1), 284–298.

Goncharuk, A.G. (2017). Wine value chains: Challenges and prospects. *Journal of Applied Management and Investments*, 6(1), 11–27.

Juchau, R. (1996). Australian wine companies: Asset reporting practices. *Australian Accountant*, 66, 32–37.

Kotler, P. (2002). *Marketing Management*, 2nd edition. Pearson Custom Publishing, Boston.

Koulayom, H. and Chopov, B. (2007). Quelles perspectives pour la recherche en comptabilité et en finance dans les entreprises évoluant dans le secteur vitivinicole en France ? Working Paper, Hal Open Sciences, hal-01147878.

Koumanakos, D.P. (2008). The effect of inventories management on firm performance. *International Journal of Productivity and Performance Management*, 57(5), 355–369.

Lebans, M. and Euske, K. (2006). A conceptual and operational delineation of performance. *Business Performance Measurement: Theory and Pratice*, 1, 65–79.

Lieberman, M.B., Helper, S., Demeester, L. (1999). The empirical determinants of inventories levels in high-volume manufacturing. *Production and Operations Management*, 8(1), 44–55.

Mathuva, D.M. (2013). Determinants of corporate inventories holdings: Evidence from a developing country. *The International Journal of Applied Economics and Finance*, 7(1), 1.

Michalski, G. (2008). Corporate inventories management with value maximization in view. *Agricultural Economics*, 54(5), 187–192.

Moccia, L. (2013). Operational research in the wine supply chain. *Information Systems and Operational Research*, 51(2), 53–63.

Muller, M. (2019). *Essentials of Inventory Management*. Harper Collins, New York.

Qian, Z. (2018). *Dynamic Linkages between Corporate Diversification Strategies and Operational Performance: An Empirical Investigation*. ProQuest, Ann Arbor.

Rodriguez, P. and Ocejo, J. (2002). Aspectos del plan general contable de las empresas del sector vitivinícola (Issues of the general accounting plan to the wine sector). *Téchnica Contable*, 54(639), 197–220.

Scherr, F.C. (1989). *Modern Working Capital Management: Text and Cases*. Prentice Hall, Hoboken.

Wild, T. (2017). *Best Practice in Inventories Management*. Routledge, London.

2

Is It Odd to Leverage Data Analytics in SMEs? A Study on Marketing and Innovation Practices

2.1. Introduction

For several decades, digital technologies have been transforming our economies and societies, affecting all sectors of activity and the daily lives of individuals. This digital transformation is disrupting norms and practices in organizations and also offers new opportunities to create value and innovation. We are seeing that firms are integrating digital technologies more into their business model and strategy to take advantage of these opportunities. During the Covid-19 crisis, this transformation accelerated, with the use of new digital tools for communication, customer relationship management and e-commerce.

Digital data is also at the heart of this transformation. Since the 2000s, the Internet has introduced new uses for individuals with mobile applications, social media and e-commerce, so much so that today every user action generates significant amounts of digital data. This phenomenon of digital data hyperproduction is called Big Data. Big Data is of great interest to firms because it is a source of valuable information regarding their customers and market. More and more firms are turning to Data Science and data analytics to improve their customer knowledge, decision-making and performance. Marketing is one of the areas where data analytics has the most performance prospects.

Chapter written by Théo JUSTY, Estelle PELLEGRIN-BOUCHER, Julien GRANATA and Denis LESCOP.

However, research on Big Data and data analytics in small- and medium-sized enterprises (SMEs) remains limited. From a general standpoint, little research has focused on low-tech industries, of which the wine sector is a part. In addition, the wine sector is dominated by SMEs with limited resources and singular strategies (Granata et al. 2018). Today, the challenge is to better understand which uses of data analytics can be beneficial for SMEs, for several reasons.

First, SMEs represent the majority of the French economic fabric and perceive the benefits of digital technology. According to a 2021 study by France Num, 78% of micro firms and SME owners-managers believe that digital technologies represent a real benefit for their company. Second, they are facing the challenges of Big Data and are investing increasingly in data analytics software. SMEs in the wine sector have to deal with the digitization of their business management or their agricultural operations (Cannas 2023). In the context of the fourth industrial revolution characterized by a strong digitization of industrial and agricultural sectors, data analytics help SMEs to effectively use their financial and marketing resources and strengthen the resilience of organizations in the face of economic upheavals. For wine SMEs, the challenge is to seize this digital opportunity and identify strategic uses of data analytics in their organization.

In this chapter, we will specifically address the case of industrial SME in the wine sector. Data analytics is of great interest to SMEs, which are becoming aware of the business value potential of Big Data. The wine sector has been undertaking a real digitization of its operations in recent years. It now wants to develop data analytics-driven business management practices (Cannas 2023). Wine SMEs also have the ambition to improve their innovation and marketing capabilities, especially since the Covid-19 crisis.

This chapter presents an in-depth literature review on Big Data and data analytics in SMEs, as well as the results of an exploratory study carried out in 2021 in collaboration with the Vinseo corporate network. The latter brings together more than a hundred SMEs in manufacturing and agricultural sectors. This corporate network links many industry players, regional public institutions and research laboratories. Our results show that data analytics allows SMEs in the wine sector to improve customer knowledge, market segmentation, marketing strategy, marketing communications personalization and online customer experience. We also explored the positive influence of data analytics on firms' ability to make financial predictions and adapt to market turbulence. Therefore, we contribute to better understanding data analytics practices in marketing and innovation, in an SME context.

In section 2.2, we will define the concepts of digital transformation and Big Data to understand their impact on business management. In section 2.3, we will show that data analytics adoption is within the reach of SMEs, and will analyze the role of data analytics in the management of wine SMEs. Finally, in section 2.4, we will study the performance prospects of data analytics in the marketing and innovation operations of SMEs.

2.2. Major digital trends among SMEs

2.2.1. *Digital transformation and its impact on business management*

For several years, digital transformation has become a key topic in research in strategy, marketing, and information systems, as well as in firms. Digitization refers to the process of transforming a firm's business, processes, skills and business model to take full advantage of the benefits and opportunities brought by digital technologies, following a strategically orchestrated approach. Digital transformation typically involves digitalizing processes to optimize efficiency and performance, and innovation to improve existing products and services through digital technologies. We can group digital technologies under the acronym SMACIT, namely technologies for social media, mobile applications, data analytics, cloud-based and Internet of Things (Vial 2019).

At an advanced level, digital transformation drives profound changes in a firm's organization as the use of digital technologies intensifies (Majchrzak et al. 2016). Recent studies have helped improve our understanding of these digital-related changes. At an organizational level, research shows that firms need to take a broader view of the matter to succeed in their digital transformation. Digitalizing a firm's organization is not just about implementing digital business management software, such as an ERP or CRM, or building a customer database. Digital technologies themselves are only part of the complex puzzle that firms need to solve in order to successfully digitalize and remain competitive. Digital transformation is a holistic approach that must take place in every person's occupation and at every level of the organization. Firms need to undertake changes in their organizational structure, processes, information systems and culture (Bharadwaj et al. 2013).

Digital transformation also involves that firms make changes to their strategic vision. The digital business strategy is a master plan that helps firms to design a strategic vision that integrates digital technologies and manages transformations related to the realization of new innovative and differentiating operations (Mithas et al. 2013). Therefore, the benefits of digital transformation are all the more significant when it aligns with the firm's strategy. Strategy and digital alignment is a continuous process of adaptation and change that requires feedback between the firm's needs, market developments and digital technologies (Canhoto et al. 2021). Aligning the digital strategy is a real challenge for today's business managers. Nevertheless, if this alignment is respected, digital technologies can improve marketing and sales operations, help promote the brand, optimize knowledge management and customer relationship management, promote product and managerial innovation, or support business growth.

More generally, digitization impacts all operations in the business value chain. Its importance has gradually increased in firms as digital technologies for management, accounting, communication and marketing have multiplied. Digitization also affects individuals' consumption patterns, as well as firms' customer journeys and sales channels. Activities such as marketing, sales and customer relationship management are increasingly digitalized to adapt to new consumer uses on the Internet (Chen et al. 2012).

Recently, the Covid-19 crisis has acted as an accelerator for a firm's digital transformation. The health crisis has not only disrupted individuals' consumption behaviors, but also firms' commercial operations. Firms were forced to precipitate the digital transformation of work, strategy and operations management (Papadopoulos et al. 2020). A McKinsey Institute study published in October 2020 showed that the Covid-19 crisis accelerated the digital transformation of firms in Europe by about seven years, an increase of three years compared to pre-crisis forecasts. At an organizational level, remote work pushed firms to equip themselves with videoconferencing software for a subscription, as well as cloud-based storage space. In business-to-customers industries, 55% of customer interactions are now dematerialized in Europe, as firms have been forced to rapidly digitalize their product and service offerings in response to the mass migration of consumers to online sales channels.

2.2.2. *Big Data and data analytics technologies*

Since the 1990s, the world has been digitalizing on a large scale with the development of the Internet. Since the 2010s, many studies have focused on the impact of the Internet on the global production of digital data, and for good reason: the Internet has revolutionized Internet users' utilization, both socially and professionally, with the development of social media such as Facebook, Instagram and LinkedIn. These platforms bring together billions of users and make it easy to create or consume content online. In addition, e-commerce has transformed the habits of consuming goods and services on the Internet. Today, individuals consent to share personal data in return for online services, such as subscribing to a service or placing an order. Finally, 4G and devices such as smartphones, tablets, smartwatches and their app ecosystems, allow individuals to move with their digital environment to any location. In the manufacturing industry, firms are equipping themselves with smart production machines and digital business management software to oversee their production and supply chain. Today is the era of the Internet of Things and Industry 4.0 (Chen et al. 2012).

Therefore, all the daily actions and interactions of users in virtual and physical spaces generate massive amounts of digital data. In 2010, Eric Schmidt, former CEO of Google, told at the Atmosphere Convention: "There were 5 Exabytes of information created between the dawn of civilization through 2003, but that much information is now created every 2 days." This figure highlights the extent of the Big Data phenomenon in our society today.

Big Data generally refers to the phenomenon of hyperproduction of digital data on a global scale caused by the digitization of our lifestyles and production, and whose complexity exceeds the processing capabilities of traditional information systems (Manyika et al. 2011). Big Data is based on three fundamental dimensions called the "3 Vs", namely volume, velocity and variety. These dimensions illustrate the abundance of digital data volumes produced, the velocity at which they are produced, and the wide variety of data sources and formats. Gradually, other characteristic dimensions were identified as veracity and value. Value refers to the economic potential of digital data related to the business value of the information contained in the databases and the performance prospects for firms (Wamba et al. 2015).

In order to study the societal and managerial challenges of Big Data, a new field of research has developed, called Data Science. It is a knowledge system composed of new theories, methods and technologies to transform digital data into useful information applicable to business management. Data Science is built around a set of interdependent and complementary technologies, such as Knowledge Management, Data Mining, Machine Learning, Deep Learning, Business Intelligence, automation, etc. There are other research sub-fields such as artificial intelligence (AI) that contribute to the improvement of methods and technologies for data analytics. For example, AI focuses on developing algorithms to solve complex or time-consuming problems by simulating human cognitive abilities (Xu et al. 2021). All of these methodologies and technologies can be used to analyze digital data. Data analytics is the process of collecting, analyzing, visualizing and transforming digital data into information relevant to decision-making, business strategy and business management (Manyika et al. 2011). This process is sometimes called "data valorization" by practitioners because it is about turning digital data into valuable information for business management.

In practice, analyzing digital data involves analytical software also called Data Analytics. It is software dedicated to the collection, storage, sharing, analysis and visualization of digital data. These tools can take the form of software or extensions integrated into an ERP or CRM. These analytical tools are also available as software as a service. There are three types of analytical tools: descriptive, predictive and prescriptive. Descriptive data analytics report periodically on an organization's key performance indicators. Predictive data analytics involve identifying or extrapolating short- and long-term trends from historical data. Finally, prescriptive data analytics serve as decision aids and formulate operational suggestions. They depend on the activity and use case of the organization (Chen et al. 2012).

Big Data began to generate real interest among companies after McKinsey Institute published a series of studies in the 2010s, including "Big Data: The Next Frontier for Innovation, Competition, and Productivity" by Manyika et al. (2011). Since then, Big Data has been a subject of great importance, both in academic research and in firms. This is for good reason, as companies are realizing that digital data about their customers or their operations can be exploited for the benefit of their performance. Indeed, many studies highlight the positive impact of data analytics on decision-making, process optimization, business strategy and business models. Today, there are a large number of applications in financial management, supply chain, human resources, innovation, marketing and sales.

Marketing seems to be a promising field for data analytics, as it is one of the most digital, data-producing functions of the company. Indeed, data analytics make it possible to deepen firms' customer knowledge. Data analytics is also involved in improving marketing strategy and operations, and personalizing the customer experience. Today, research is increasingly focused on Big Data and data analytics in SME operations.

2.3. SME craze for Big Data: applicability in the manufacturing and agricultural sectors

2.3.1. *Challenges around digital data in SMEs: the wine sector case*

Data analytics is now a strategic issue for firms in all sectors of activity. Leveraging data analytics opens up new avenues for innovation and efficiency, particularly for small- and medium-sized businesses. Indeed, SMEs are well aware of the benefits and opportunities associated with digitalization and Big Data, especially since the Covid-19 crisis. Today, more than ever, SMEs are looking to innovate, take advantage of digital technologies and strengthen their ability to adapt to crises, which data analytics could enable them to do. Indeed, the European Commission acknowledges the economic potential of digital data and has incorporated it into its strategy. In 2022, they proposed a law, the "Data Act", aimed at enhancing the utilization of industrial digital data in small- and medium-sized enterprises.

Research on practices and performance prospects of data analytics in SMEs is still limited. Some recent studies, such as that of Maroufkhani et al. (2020), have begun to study the impact of data analytics on SME performance. However, there is a need to improve our knowledge on data analytics in SMEs, for several reasons. First, SMEs with fewer than 250 employees and an annual turnover of less than 50 million euros represent 98% of firms in France and Europe. These are a major player in the French and European economy and contribute up to 40% to the total added value created in France. SMEs are also the leading source of employment in France with 6.3 million employees. Finally, SMEs invest approximately 18% of their turnover per year in R&D and innovation (CEDEF 2018). Due to their significance in the economy, we felt it was necessary to further study the uses of data analytics in SMEs, in order to better support them in the adoption data analytics.

Second, SMEs are facing the challenges of digital transformation. For several years, they have been investing increasingly in digitization for their organization and in digital management solutions, such as ERP, CRM and Business Intelligence. We are also seeing a growing craze for Big Data, AI and process automation in SMEs. Since the Covid-19 crisis, SMEs have drastically accelerated their digitization, in order to maintain their operations, the link with their employees and customers, and their competitiveness. Today, SMEs are recognizing the tangible benefits of digital technologies, with data analytics posing as one of their key technological goals. They are definitely looking for new strategies or digital technologies to reinforce their competitive advantage in their market (Cannas 2023).

Finally, firms, including SMEs, collect large amounts of data about their operations and customers through digital business management software. They fully understand that the smart utilization of digital data is a significant driver for performance and innovation. Therefore, they seek to understand how to successfully adopt data analytics into their organization and how to develop a data-driven strategy. Recent studies have indeed highlighted a strong link between data analytics and the financial and marketing performance of SMEs. In addition, it is important to identify the obstacles they face, in order to better support them in data analytics adoption. This is also relevant to public and private organizations that offer guidance and consulting services to SMEs (Maroufkhani et al. 2020).

We find these challenges in SMEs in many so-called low-tech sectors. These are characterized by the use of digital technologies dedicated to business management or production that are less sophisticated, less expensive and technically more accessible. They oppose so-called high-tech sectors that invest heavily in advanced technologies to meet the needs of a niche market. Low-tech sectors are vectors of major innovations and are able to exploit high-tech technologies (Czarnitzki and Thorwarth 2012). Data analytics is a major technological innovation for SMEs in a low-tech sector such as the manufacturing and agricultural industries. For a long time, digital technologies were mainly present in R&D, engineering and production. With Industry 4.0, the development of smart production devices and precision agriculture, SMEs have increased their use of industrial and agricultural digital data as part of their production operations. At the same time, industrial SME and software editors in the manufacturing and agricultural sectors have become aware of the importance of collecting and analyzing their financial, customer and business data. The Covid-19 crisis has obviously pushed SMEs in these sectors to leverage digital technologies and their digital data to cope with the interruption of their sales and supply chain operations. Since the crisis, progressively more manufacturing and agricultural SMEs want to invest more in data analytics to improve their customer knowledge, develop an effective and dynamic strategy, or innovate to offer new digital products and services.

This trend seems to be confirmed in these sectors where there is significant digital transformation. Unfortunately, research on low-tech industries, such as the wine sector, is still very limited. However, the wine market is particularly attractive for the development of innovative solutions (Granata et al. 2019), such as data analytics that has the potential to improve the efficiency of production processes and the connection between firms and customers (Spadoni et al. 2019). In a sector as competitive as this, companies that invest in data analytics for their business management or strategy, improve their agility and responsiveness to market changes (Cannas 2023). This competitive advantage is significant for wine SMEs, whose limited resources lead them to develop innovative solutions (Granata et al. 2018).

2.3.2. *Digital data and analytical tools in wine SMEs*

In 2021, we began our research on Big Data and SMEs by conducting an exploratory study. One of our objectives was to better understand the practices in data analytics in SME business management in the manufacturing and agricultural sector. For this purpose, we collaborated with the Vinseo corporate network. Located in the Occitanie region of France, this corporate network brings together more than a hundred SMEs in goods and services in the manufacturing and agricultural industry sectors. Vinseo also has many research and teaching partners, including the University of Montpellier, the Agricultural Institute of Montpellier and the French Institute of Wine.

We targeted firms that meet the criteria of an SME, namely less than 250 employees and less than 50 million euros in turnover. We conducted a total of 15 semi-structured interviews with 13 SMEs specializing in manufacturing, industrial equipment distribution, business-to-business service provision and software editing. Our sample includes owners-managers, sales and marketing managers, and R&D project managers.

Initial results show that the majority of manufacturing and agricultural SMEs define data analytics as a tool to assess and understand in business management. Indeed, firms in these sectors collect digital data for various uses, namely descriptive, predictive and prescriptive practices. In descriptive practices, data analytics make it possible to accurately measure, understand and track a firm's activity. Descriptive data analytics give SMEs a better view of the current state of the business. In predictive practices, data analytics make it possible to make financial or sales predictions. These are statistical analyses to identify or anticipate short- or long-term trends. In prescriptive practices, data analytics make it possible to formulate operational suggestions and identify areas for process optimization.

We also find that the digital data in manufacturing and agricultural SMEs comes from many sources. These are both internal and external sources. Internally, these sources include digital data, such as financial, human resources, production and business data. They are generated by business management software, such as HRIS, ERP and CRM. It should be noted that many SMEs in the sector truly began digitalizing paper archives before and during the health crisis, in order to create new financial and customer databases, and to have a complete history in their business management software. Externally, these sources include digital data collected from their websites, social media and e-commerce platforms. It is also open data or data generated by smart devices.

For digital data storage, many SMEs have their own physical servers. The IS department is generally responsible for maintaining the servers. However, progressively more SMEs in the manufacturing and agricultural sectors are acquiring cloud-based storage space. Cloud technologies have lower maintenance costs and allow for more flexible and efficient access to digital data. SMEs are also increasingly using a service provider for designing and maintaining their information system architecture. From a general point of view, digital data storage is a crucial issue for SMEs. They are aware of issues related to the protection of personal data and cybersecurity. According to the European Commission's "Cybersecurity" barometer, about 20% of French SMEs were victims of cyberattacks in 2021. They have a lot of digital data about their firm, as well as their customers. The value and confidentiality of this information usually requires a great deal of care. The SMEs interviewed explained that it was essential to invest in secure digital technologies that were compatible with the General Data Protection Regulation (GDPR). Manufacturing and agricultural SMEs are also well aware of the environmental issues in their sector. Since data storage is an energy-intensive operation, they seek cloud services with a lower environmental impact, in order to achieve their CSR and energy efficiency objectives.

For data analytics, the nature and sophistication of these software in SMEs vary according to the uses and strategy of each company. The size of the structure is not necessarily a determining factor. We have noticed that the sophistication level of data analytics in SMEs varies according to digital data, when these are part of the service offer, according to management, whether it relies on digital data for its strategy or operations, according to its level of digitization, and according to its skills. Therefore, some SMEs use Excel or collaborative tools to analyze their digital data; others analyze their digital data thanks to their ERP, CRM and Business Intelligence software or develop their own algorithms for data analysis and prediction. In any case, our research shows that manufacturing and agricultural SMEs are in the midst of a transition towards more data analytics-based business management, particularly in marketing and innovation.

2.4. Uses of data analytics in marketing and innovation in SMEs

2.4.1. *Improving customer knowledge, marketing strategy and personalization*

Digital technologies offer many ways to collect customer information. Customer knowledge is rightly considered one of a firm's most valuable assets. Many digital business management software today allows firms, including SMEs, to actively create and use customer data, in order to best meet the needs and developments of their market. Digital data thus plays an important role in improving customer knowledge (Gupta et al. 2021).

In the manufacturing and agricultural sectors, we have also found that SMEs are increasingly collecting digital data on their customers, as well as on the use that these customers have of their products or services. They invest in creating databases of basic information on customers, projects or orders in progress, transaction history, revenue generated, marketing campaigns, etc. This digital data is collected through their ERP, CRM, digital after-sales service platforms or other analytical software. At the same time, firms in these sectors collect a lot of customer data online. Many SMEs created online services before and during the Covid-19 crisis. Search engines, social media and e-commerce platforms allow them to track customers' online activity. By analyzing metrics on traffic, click-through rate, sales and marketing operations, data analytics offer a comprehensive view of online individuals' purchasing behaviors.

With digitization, SMEs are thus observing an improvement in customer knowledge management, namely the process of identifying, acquiring and using customer information. This data allows SMEs to better identify their customers' characteristics, needs and expectations. The idea is to know your market better to better serve your customers. It is possible to better understand the buying process and customer journey, as well as identify buying trends in a targeted market. SMEs consider that improving customer knowledge is a real differentiator for their firms.

Better customer knowledge would also improve marketing strategy development, and, more specifically, segmentation, targeting and positioning. To improve market segmentation, some SMEs analyze their customer databases to determine target customer profiles and identify the most profitable customer groups. They sometimes adopt a different marketing and sales strategy depending on targeted segments. Also, some SMEs rely on their customer data to improve the targeting of their marketing operations. With more in-depth knowledge, SMEs manage to carry out more effective and personalized operations with customers who are most likely to be interested in their offer. Finally, collecting data on customer expectations allows SMEs to consolidate their brand positioning. This information sometimes helps these companies to build the image of their product or service, and to position their offer according to a particular purchasing criterion.

Personalizing marketing communications or the online customer experience brings a lot of value to customer relationship management. Data analytics can help firms like SMEs personalize the digital environment they offer their customers. The challenge is understanding customer expectations and adapting to their uses in order to develop a close relationship with them (Chen et al. 2012).

With the Covid-19 crisis, many SMEs, such as producers, cooperatives and suppliers, created websites and developed online services or e-commerce platforms to maintain their business operations. They were keen to create a pleasant and efficient online customer experience, as well as well-designed product pages. Optimization and personalization of a website is recognized as a key success factor in online sales for good reason. Some wine SMEs are able to improve the online customer experience by analyzing data on best-selling products, the most or least visited web pages, and the customer journey. They make necessary changes and create an optimized website. Some marketing or e-commerce managers have even integrated product recommendation systems into the website. The goal is to analyze customers' past online behavior and present product information that matches customers' preferences.

Customer data also helps SMEs to personalize their marketing and sales operations. Firms can now personalize their communications using a CRM or a digital marketing piece of software. Some SMEs identify specific prospects or customers and create personalized marketing messages, for example, to present an offer, make an appointment or invite people to a trade show. Our research shows that SMEs have much higher opening and conversion rates with personalized marketing and sales operations, thanks to customer data analytics.

2.4.2. *Developing incremental and radical innovation through data analytics*

Data analytics also play a role in designing innovative products and services. We usually distinguish between two main types of innovation: incremental innovation and radical innovation. Incremental innovation aims to make minor changes to a product or service. Radical innovation is the development and introduction of an entirely new product or service to the market. These are considered essential to gaining and maintaining a competitive advantage. Recent studies show that data analytics would have a positive impact on these forms of innovation in SMEs (Maes and Sels 2014).

This is a performance perspective that SMEs are well aware of. In terms of incremental innovation, small firms collect a lot of customer and industrial data, and wish to leverage it to improve their innovation capabilities. They collect this data in contact with their customers, on online discussion forums, and as part of their satisfaction surveys or after-sales service. Their objective is to continuously improve their products or services, in order to meet the needs of their market ever more precisely, and to increase the added value of their offer. They forward these suggestions to their R&D department, with whom they make changes to existing products or services.

Regarding radical innovation, SMEs are increasingly integrating data analytics into their business model. They see data analytics as a way to develop innovative services and monetize their agricultural and industrial databases. Some SMEs develop innovative services, such as data-driven consulting services or data analytics-based services in addition to their offer and smart wine tourism services. This type of innovative services around data analytics has greatly helped some firms in these sectors during (and after) the health crisis to maintain their operations and competitiveness. Other SMEs have a history of agricultural, industrial and oenological data over several decades and regularly feed these databases. They condition these volumes of digital data into a structured database, which they integrate into their product offering. They give access to this information to their customers for a fixed price or through a subscription.

Other forms of innovation associated with data analytics in SMEs can also be identified. They analyze, for example, their data to optimize their industrial and business processes. Others rely heavily on digital data as part of their business management or strategy, so we can call it managerial innovation. In short, data analytics seems to help many firms in these sectors to innovate at various levels and maintain their competitiveness.

2.4.3. *Prediction and crisis adaptability*

Predictions are crucial for managing and adjusting firms' resource allocation, strategy, marketing and production capabilities. The challenge for firms is to build an actionable database to make predictions and study trends in the market. The accessibility of financial, industrial and business data has been enhanced by the advent of new digital business management software. An informed use of this data makes it possible to improve the forecast accuracy in firms. Some studies show that data analytics thus facilitates improved decision-making, lessens uncertainty and introduces better risk management (Schoenherr and Speier-Pero 2015).

Historically, forecasts are important due to increased uncertainty in markets and the need to deal with limited resources among SMEs. Today, the Covid-19 crisis has had significant consequences on the operations of SMEs in the manufacturing and agricultural sectors, and on their customers.

We have noticed that a growing number of owners-managers, as well as financial and business executives, are increasingly gathering data related to their operations to make more accurate forecasts. They also consider weather, seasonality, political decisions and possible industrial shortages in order to anticipate a change in demand in their market. Faced with an uncertain and even turbulent environment, SMEs want to be serene in their decision-making and cautious in managing their costs and risks. Therefore, SMEs can benefit from accurate business, industrial and financial predictions thanks to data analytics in order to better anticipate the evolutions of their operations.

Finally, data analytics can improve SMEs' ability to detect changes, assimilate information and quickly reconfigure their strategy and supply chain. Many manufacturing and agricultural SMEs seek to improve their responsiveness and ability to adapt to changes in their market through data analytics. Some of them see a real advantage in identifying these developments through data analytics, as their shorter decision-making processes and communication channels allow them to react quickly and gain dynamism in their operations.

2.5. Conclusion

For several decades now, two major trends have been emerging as the use of the Internet progresses in firms. On the one hand, digital transformation is driving profound changes in trades, processes and strategy, with the integration of new business management software and the development of online services. On the other hand, Big Data is taking an increasingly important place in the management and strategy of firms. With the proliferation of Internet users, online services and digital business management software, firms are collecting massive amounts of information about their customers and operations. The acceleration of these two phenomena has been particularly noticeable in the wake of the Covid-19 crisis. Today, SMEs, especially in the manufacturing, agricultural and wine sectors, see data analytics as an opportunity to improve their business management, offer new digital services and strengthen their resilience in the face of crises.

Our research shows that SMEs in these sectors have various practices in data analytics from a business management standpoint. Thanks to a study of 13 SMEs in the manufacturing, agricultural and wine sectors, we know that data analytics play an important role in the management of these small firms. Data analytics allow SMEs in these sectors to improve customer knowledge, marketing strategy, personalization of their communications, as well as to innovate on their products and optimize their prediction and adaptation capabilities. It is also important to emphasize that data analytics is a complex process that requires mobilizing a large number of resources for SMEs. This research helps to improve our understanding of digital transformation, Big Data and the practices in data analytics in SMEs, as well as better understand the performance prospects of data analytics in SMEs, through the case of the manufacturing, agricultural and wine sectors.

However, implementing and using data analytics can be a real challenge for SMEs. Some SMEs admit that this requires the integration of new resources and the abandonment of long-standing management practices. They must acquire new digital skills to integrate data analytics into their existing information systems architecture and to use it as part of their business management. They must also identify value-creating applications of data analytics and develop an organizational culture that promotes the use of digital technologies in their organization. To grasp the business value of data analytics, SMEs need to understand how to overcome these challenges, as well as the business value potential of these technologies. However, recent research on Big Data and SMEs focuses more on the impact of data analytics on business performance than on identifying organizational challenges associated with their operation (Maroufkhani et al. 2020). Future research could be conducted to identify obstacles to data analytics in SMEs, in order to provide concrete managerial recommendations on data analytics adoption.

2.6. References

Bharadwaj, A., El Sawy, O.A., Pavlou, P.A., Venkatraman, N. (2013). Digital business strategy: Toward a next generation of insights. *MIS Quarterly*, 37(2), 471–482.

Canhoto, A.I., Quinton, S., Pera, R., Molinillo, S., Simkin, L. (2021). Digital strategy aligning in SMEs: A dynamic capabilities perspective. *The Journal of Strategic Information Systems*, 30(3), 101682.

Cannas, R. (2023). Exploring digital transformation and dynamic capabilities in agrifood SMEs. *Journal of Small Business Management*, 61(4), 1611–1637.

CEDEF (2018). Où trouver les chiffres clés des PME ? Ministère de l'économie, des finances et de la relance [Online]. Available at: https://www.economie.gouv.fr/cedef/chiffres-cles-des-pme [Accessed July 2022].

Chen, H., Chiang, R.H.L., Storey, V.C. (2012). Business intelligence and analytics: From big data to big impact. *MIS Quarterly*, 36(4), 1165–1188.

Czarnitzki, D. and Thorwarth, S. (2012). Productivity effects of basic research in low-tech and high-tech industries. *Research Policy*, 41(9), 1555–1564.

Granata, J., Lasch, F., Le Roy, F., Dana, L.P. (2018). How do micro-firms manage coopetition? A study of the wine sector in France. *International Small Business Journal*, 36(3), 331–355.

Granata, J., Aytaç, B., Roubaud, D. (2019). Innovation developments in the wine industry: A journey from the amphorae of old to the California wine cluster. *International Journal of Entrepreneurship and Small Business*, 36(3), 249–255.

Gupta, S., Justy, T., Kamboj, S., Kumar, A., Kristoffersen, E. (2021). Big data and firm marketing performance: Findings from knowledge-based view. *Technological Forecasting and Social Change*, 171, 120986.

Maes, J. and Sels, L. (2014). SMEs' radical product innovation: The role of internally and externally oriented knowledge capabilities. *Journal of Small Business Management*, 52(1), 141–163.

Majchrzak, A., Markus, M.L., Wareham, J. (2016). Designing for digital transformation: Lessons for information systems research from the study of ICT and societal challenges. *MIS Quarterly*, 40(2), 267–278.

Manyika, J., Chui, M., Brown, B., Bughin, J., Dobbs, R., Roxburgh, C., Byers, A.H. (2011). *Big Data: The Next Frontier for Innovation, Competition, and Productivity*. McKinsey Global Institute, New York.

Maroufkhani, P., Tseng, M.L., Iranmanesh, M., Ismail, W.K.W., Khalid, H. (2020). Big data analytics adoption: Determinants and performances among small to medium-sized enterprises. *International Journal of Information Management*, 54, 102190.

Mithas, S., Tafti, A., Mitchell, W. (2013). How a firm's competitive environment and digital strategic posture influence digital business strategy. *MIS Quarterly*, 37(2), 511–536.

Papadopoulos, T., Baltas, K.N., Balta, M.E. (2020). The use of digital technologies by small and medium enterprises during COVID-19: Implications for theory and practice. *International Journal of Information Management*, 55, 102192.

Schoenherr, T. and Speier-Pero, C. (2015). Data science, predictive analytics, and big data in supply chain management: Current state and future potential. *Journal of Business Logistics*, 36(1).

Spadoni, R., Nanetti, M., Bondanese, A., Rivaroli, S. (2019). Innovative solutions for the wine sector: The role of startups. *Wine Economics and Policy*, 8(2), 165–170.

Vial, G. (2019). Understanding digital transformation: A review and a research agenda. *The Journal of Strategic Information Systems*, 28(2), 118–144.

Wamba, S.F., Akter, S., Edwards, A., Chopin, G., Gnanzou, D. (2015). How 'big data' can make big impact: Findings from a systematic review and a longitudinal case study. *International Journal of Production Economics*, 165, 234–246.

Xu, Z., Tang, N., Xu, C., Cheng, X. (2021). Data science: Connotation, methods, technologies, and development. *Data Science and Management*, 1(1), 32–37.

3

What is the Cost of a Hectare of Vines? Financial Value Versus Emotional Value: The Petrus Case

3.1. Introduction

The dichotomy between value and price is not new, and price formation for real assets is increasingly complex. Indeed, the outcome is not only a result of a combination of various values, which can sometimes develop in conflicting ways, but also of interactions among different stakeholders such as buyers, sellers, evaluators and advisors. Ultimately, it is a matter of conviction and perception.

Financial crises that have taken place for more than a decade have helped to shake many certainties, including how the value of a company can be correctly apprehended by traditional valuation methods. These methods aim to provide a universal and identical financial value of a company, regardless of the potential investor or transferor. On the other hand, the strategic value reflects how an investor perceives the company's worth, based on their strategic evaluation of the target company, which includes an emotional component.

So, the value of a company does not exist; there are as many strategic values as there are potential investors. The distinction between financial value and strategic value thus provides a better understanding of the difference between value and price.

It is particularly rewarding to illustrate this problem by studying the transfer or sale of vineyards; while their economic value is often low, the price paid to acquire them is skyrocketing, since these farms are high in emotional appeal.

Chapter written by Gérard HIRIGOYEN and Pascale WEBER.

In this chapter, we focus on an emblematic case, namely the sale of 20% of Petrus' capital to a Colombian billionaire in November 2016, which would make it the most expensive hectare of vineyard in the world if the valuation of this minority stake were to be confirmed at more than 200 million euros. Thus, it raises questions about the buyer's motives.

Since its inception, the Petrus case has been a textbook case for marketing students. This chapter will also give it a strategic and financial dimension.

The Petrus vineyard consists of 11.3 hectares. The original plots are located on the highest area of the Pomerol appellation, at about 40 m above sea level, on a plateau made of blue clays. The average age of the vines is about 40 years. The estate produces an average of 30,000 bottles, depending on the year.

While it already existed at the end of the 18th century, owned by the Arnaud family since 1770, it was known as Petrus-Arnaud. The origin of its name is uncertain; it would refer to St. Peter or the first Pope Petrus.

Jean-Pierre Moueix, the merchant and producer from Libournais, after having been the exclusive distributor of Petrus wines, eventually became the owner in 1969, following the death of Mrs. Edmonde Loubat in 1961 and the redemption of her nephews' shares. He doubled the property in size by acquiring 5 hectares at Château Gazin.

Petrus is arguably the most famous wine in the world, albeit not in any ranking. Unlike other famous Bordeaux wines, such as Château Latour, Château Margaux, Château Haut-Brion and Château Ausone, Petrus is not a historical Bordeaux wine and, above all, its fame is recent, since it can be dated from 1945, or even the early 1970s, when Jean-Pierre Moueix became the head of the estate, when he introduced Petrus to the table of President Kennedy at the White House and gave birth to his legend.

Petrus is labeled without the word "château" (and without an accent), which sets it apart from other wines from the Bordeaux region. However, it should be noted that very old labels have the name "Château Pétrus" with an accent and mention "*mis en bouteille au Château*" (bottled at the Château).

After the death of Jean-Pierre Moueix in 2003, his son Jean-François, president of the trading company Duclot, inherited it. Since then, his son Jean, president of the family holding company Videlot, which owns Petrus and the trading company Duclot, has taken over. Apart from Petrus, Videlot has only two minority shareholdings, one with Château Calon-Ségur 5% and the other with Château Montrose (2%), two flagships of Saint-Estèphe.

Continuity is also ensured on the side of the cellars by the Berrouet family. From 1964 to 1967, Jean-Claude Berrouet was the oenologist in charge of winemaking at the Moueix properties and the one who talentedly sublimated the production of merlot and its exceptional terroir. Since the 2008 vintage, his son Olivier has been performing this role.

The acquisition of a minority stake in Petrus and its exceptional valuation presents an ideal case study. This scenario not only deepens our understanding of the factors influencing the purchase prices of vineyards but also offers a comprehensive view of various valuation approaches and perspectives.

However, the issue under examination extends beyond the specific example chosen. It is in fact the authors' questioning of the explanation of discrepancies observed, on the one hand, between appellations, concerning evolution rates and value levels, and, on the other hand, within the same appellation, between the exchange price of a hectare of vineyard of certain properties and the historical average value of their appellation of origin.

These questions make it possible to introduce the role of emotional value in price formulation.

3.2. Valuation of Petrus

3.2.1. *A record value per hectare*

The information was revealed in the first days of September 2018, after being hushed for over a year: 20% of Petrus' capital was sold to Alejandro Santo Domingo, an investor of Colombian origin. The exact amount of the transaction has not been disclosed, but the Pomerol property (11.5 hectares) would have been valued at more than 1 billion euros. If confirmed, this value would make the Petrus hectare the most expensive ever sold.

The amount transacted was approximately 200 million euros, equivalent to 87 million euros per hectare, or more than 40 times the price of its appellation per hectare. Thus, it embodies the partial purchase operation of this largest size in the wine world.

For comparison:

– The Clos de Tart, a jewel of Burgundy (7.53 hectares) was bought by François Pinault for 280 million euros in 2018, or 30.14 million euros per hectare, making it the most expensive transaction in Burgundy history.

– The Bonneau du Martray estate (200 years of family history), for lack of heir, passed under the American flag in 2017: Stanley Kroenke acquired 80% of the capital of this Burgundy star for more than 100 million euros.

– The price of a hectare of vineyard in Champagne was 1.04 million euros in 2021, a decrease of 5.6% compared to 2020.

This acquisition of a minority stake raises several questions.

3.2.2. *Acquisition of a stake*

Opening capital by taking a minority stake is a structuring decision in a family or patrimonial business that leads to a new risk sharing: those related to investor quality and shareholder cohabitation.

Unlike managerial companies, risk analysis for these companies cannot be carried out solely on the basis of the profitability–risk pair. Long-term challenges for companies and their managers must also be considered in deciding to open capital. Therefore, the following quintet should be considered: profitability, risk, liquidity, company sustainability, and stability of family and/or founder control (Hirigoyen and Weber 2019).

From the point of view of the target company, the question is whether this equity investment leads to a reformulation of the governance model, or even to its questioning. For a company's founder and/or their family, it is crucial to grasp the implications of allowing a minority stakeholder to invest in their business. This includes understanding the impact on strategic decision-making control, particularly when it comes to determining how value is created and distributed.

There are numerous instances of founders who end up regretting their decision to open up their company's equity. In fact, some even consider delisting their company from the stock exchange, arguing excessive pressure from minority investors on the company's strategic direction.

The ideal partner is one who aligns with the values of the historical leaders, commits to a long-term entrepreneurial culture rather than a strictly financial one, prioritizes wealth accumulation over profit distribution, and shares the same goals of company sustainability and maintaining the control exerted by the founder or their heirs. In other words, a partner who resonates with the *affectio societatis* of the founding or family shareholders. This seems to be the case with Santo Domingo's noncontrolling stake in Petrus.

However, it is necessary to question stakeholders' actions in more depth.

3.3. Stakeholders' motivations

3.3.1. *On the transferor's side*

Several hypotheses can be considered:

– potential financial difficulties within the operation: recapitalization by bringing in new investors who can inject liquidity and thus facilitate a company's revival? This is apparently not the Moueix family's situation;

– investments shared by two families, who also know each other well. This resulted in a first step, by taking a stake in Petrus's equity;

– backed by a major player whose distribution expertise can be valuable, including on the counterfeiting issues they have a perfect grasp of. Indeed, Petrus is regularly confronted with these issues which require heavy investments in traceability and authentication;

– benefit from the distribution channels of AB InBev beer, owned by the family, and share know-how;

– inheritance tax financing.

For Jean Moueix, the reason, even if it is somewhat surprising, must be found elsewhere: "An important difference with the sales that may have taken place in Burgundy is that we did not sell Petrus. We sold a minority share of 20% with a view to longevity. It may be curious to say, but it is one of the key reasons" (interview with Jean Moueix in *Le Monde* on December 17, 2018).

In fact, it can be understood that the amount obtained by the transfer makes it possible to anticipate possible transfer problems specific to family businesses. In this type of property, it often happens that the wish of one of the co-owners or heirs to sell their shares is at the origin of a sale of the entire company, because other co-owners are not able to redeem outgoing shares. In this case, the result of the transfer of the 20% makes it possible to mitigate such an eventuality, constituting insurance on the future for the children, Jean Moueix, his sister and half-sister, and Jean-François Moueix' grandchildren.

3.3.2. *On the buyer's side*

Acquisition of a minority stake also raises questions on the buyer's side, insofar as it raises the question of share value without the right to control in an unlisted company and more generally that of "trapped capital".

The following motivations can be considered to justify such a participation:

– Anticipate profit prospects that are clearly in the long term.

– Make a symbolic and prestigious investment, because, up to 20%, it remains a minority and cannot intervene in estate governance. The purchase of a wine property is a sign of personal and professional success. It materializes an entrepreneurial and/or patrimonial success and is a means of reallocating other economic and financial assets towards value-creating operations with potential, which can also constitute safe haven investments. Such a symbolic investment also includes a "spiritual" dimension, as defined by Holbrook (1994), insofar as it allows communion with the other.

– Acquire psychological property and associated rights. This is expressed by the following equality: rights of family shareholders = rights related to property + rights related to psychological property (Hirigoyen 2014). The use of this concept makes it possible to understand why, in family businesses, family shareholders behave as if they hold rights greater than what is actually conferred on them by their level of ownership. Psychological property is essentially emotional and is a feeling of possession that a person has about something or someone, due to the close or familiar relationships they have with them (Pierce et al. 2001). Moreover, this psychological property has been very frequently cultivated in these people since their childhood.

– Seek emotional gains that are psychological benefits such as reputation, personal satisfaction, participation in the creation and realization of a project, company longevity, etc. They correspond to private benefits of a nonmonetary nature (Berrone et al. 2012). When compared to the emotional costs sustained by the family shareholder, which can be attributed to personal sacrifices, the burden of responsibilities, and stress, they make it possible to identify the emotional result (emotional gains–emotional costs). This result relative to the emotional value makes it possible to define the emotional performance of the family shareholder, who arbitrates the search for an immediate financial return.

3.4. Valuation dimensions

3.4.1. *Relevance of financial value*

If the value is hiding in the shadows, how do you bring it to light?

Is market value a satisfactory measure of performance when the business is not for sale? This is an objective value.

Valuation of the vineyard hectare using the usual financial methods based on yield does not allow us to understand the evolution of the prices of the vineyard hectare nor its level. We find, in fact, sometimes a valuation lower than the price, sometimes higher, qualified as goodwill or badwill, without us knowing how to clearly affect it.

Since the implementation of IFRS standards, which have "sanctified" fair value, the understanding of price formation and its differences in value have materialized. After the transaction, differences between the transaction price and the estimated value over time are monitored and must be explained, just as we try, with *goodwill*, to explain the differences between book value and price.

The impairment test will compare the carrying amount of the asset to its future profits, which come either from the continued use of the asset (use value) or from its disposal on the market (disposal price). But how to estimate the transfer price on the market when the economic value is reduced to the congruent portion? The gaps that remain, sometimes "subsidiary" and sometimes "substantial", would certainly find a basis for integrating emotional value in determining the causes of these differences.

Only the so-called "comparative" methods allow for an approach to exchange prices without explaining their sources.

In order to understand the origin of their disparities, we will therefore rightly focus on the emotional component of value to explain prices.

3.4.2. *Emotional value*

Emotional value results from behavioral bias of an emotional nature:

– which does not place us within the framework of the classic or neoclassic theory of an objective value;

– which also does not place us in the context of a *homo economicus* type rational behavior.

It is the psychological property that also explains the emotional value and which is partly at the origin of the emotional bias.

The emotional value has a different nature depending on whether the person is on the side of the transferor or the buyer. On the transferor's side, it is a socio-emotional value: an emotional value that a family derives from its position in the shareholding and management of a company. It transcends monetary and financial aspects and is based on:

– the exercise of personal authority;

– the psychological needs of belonging;

– the concern for the preservation of family values;

– the desire to perpetuate the dynasty and maintain the founder's legacy;

– the development of familiness, which is an important characteristic of companies through exchange relations with stakeholders. This capital represents all of the resources that individuals can obtain through the knowledge of other individuals, by being integrated with them into a social network or simply by being known by them, and having a good reputation. This capital is also a source of competitive advantages for companies.

This emotional value is equal to the sum of the emotional cash flows discounted at the emotional rate of return (k_e) expected by the shareholder, whose *affectio societatis* is high (Hirigoyen 2014).

On the buyer's side, it is a perceived value: the value of an asset in the eyes of a potential buyer. It corresponds to the maximum price the buyer is willing to pay to obtain said asset. It results from the comparison between expected gains and the costs or sacrifices made for the acquisition of the asset (Zellweger 2005; Hirigoyen 2014).

Elements influencing the perceived value can be objective elements (intrinsic quality of the asset) and/or subjective elements (brand image, strategic expectations, notoriety) (Aurier et al. 2004; Sumantri and Lau 2010; Morar 2013).

3.5. Price, resulting from a combined value

3.5.1. *Price and intrinsic value*

Does this price reflect the "true value" that Samuelson (1968) called the "phantom price", as well as what Keynes (1936) called "forward-looking return" when talking about the intrinsic value of an asset?

In valuating this type of asset, vineyards, there are not only calculations on yields, but also many desires, that make it possible to give a price to such a company. However, the more the desire is maintained, the higher the company is valued and the more likely it is to achieve it.

That is why investors must share a piously opportunistic belief about its future, without which no speculation and therefore no enrichment would be possible.

In 1936, J.M. Keynes had already shown in his work "The General Theory of Employment, Interest and Money" that the financial rationality of markets was based on beliefs, even the most fabulous and disruptive.

3.5.2. *Negotiated price*

Is the "price paid" agreed upon by seller and buyer an estimate of intrinsic value? It is understood that no other estimate of intrinsic value exists.

An emotional value is added or subtracted to the financial value based on discounting future flows of economic return, from the point of view of the buyer or seller.

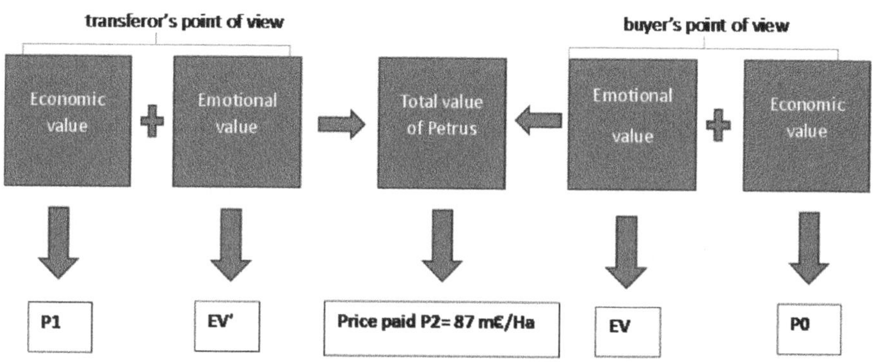

Figure 3.1. *Price, resulting from a combined value*

As with financial value, the projections of which depend on strategic visions of the potential to develop economic performance, and to be valued by everyone, emotional value will be first and foremost a subjective value. As we have seen, it is expressed by the socio-emotional flows that correspond to the satisfaction of the needs and/or expectations at an individual scale.

3.6. Consequences of such transactions

Several impacts on the wine economy can be imagined:

– The selling price of bottles can be impacted upwards, since such a record level of valuation influences the brand image of the product and its price.

– The phenomenon of transactions that are "off-market" by their amounts can become widespread, which would lead to an increase in average prices per hectare.

This would lead to prices being outbid, which could concentrate vineyards in the hands of the wealthiest, as illustrated by the transactions of Clos de Tart and the Bonneau du Martray estate in Burgundy.

– The rise in inheritance tax, calculated based on the price per hectare, could prevent the children of some winemaking families from taking over the family vineyard. Consequently, the vineyard owner may be compelled to open their capital to private investors or large corporations.

– Undermining the cohesion of family shareholding increases the desire for sale among members with low *affectio societatis*. In 2018, this was the scenario envisaged by one of the eight members of the supervisory board of Château Angelus in Saint-Émilion, who planned to sell his stake to a nonfamily third party, and who hoped to bring with him the majority of the family, an option eventually ruled out.

This family member of the supervisory board, owner of 20% of the first Grand Cru Class A of Saint-Émilion, Château Angélus, wanted to obtain the liquidity of his shares from a third party, even though he knew that a sale could not take place without the consultation and consent of the majority of the Bouard de Laforest family. According to rumors, the price asked by the seller valued the 39 hectares of Angelus at more than 500 million euros, a theoretical valuation of more than 12 million euros per hectare, which he hoped with his advice would be enough to upset other members of the family.

But the governance in place, which was eight generations strong and which had defined a mandatory and prior consultation of the family and the collection of its assent to any transfer outside it, ensured the preservation of the exclusive family character of the shareholding of this company.

The currently maintained cohesion of the majority family, loyal to ancestral commitments, thus rules out the unexpected entry of an investor not related to the family, as well as a scenario similar to Petrus with the agreed upon and planned entry of an investor who is also not a family member. The Hermès family in its time also knew how to avoid the pitfalls of an unwanted shareholding. From here, we should not take the step of concluding that family businesses should under no circumstances open up to the outside world, as there are so many abundant examples to disprove it or on the contrary confirm it.

3.7. Conclusion

The subject elaborated on in this chapter remains a hot topic; indeed, the very recent transfer of Château Lascombes (second classified growth of Margaux) at the end of 2022, to the hands of the American businessman Gaylon Lawrence, for a price estimated between 300 and 400 million euros, sets the price per hectare of the sale at nearly three times the average SAFER price of the appellation.

This example aptly contributes to the discussion in this chapter about the objectivity or subjectivity of value. This debate, which remains strikingly relevant today, historically pitted the "absolutist" viewpoint of Plato, who argued that value resides in the object itself, against Aristotle's "relativistic" perspective, which posits that value is determined by the individual who assigns it. This approach then opens the field to the consideration of psychosocial and behavioral criteria, thus complicating the identification of value components and leading to a duality of its perception, distinguishing financial value and emotional value and finally making it possible to explain the formation of the price.

This chapter is a contribution to the knowledge of the specificity of the value issue and, more generally, the evaluation of family businesses, especially unlisted ones. This problem was illustrated thanks to Petrus and the study of the sale of a minority stake to the billionaire Santo Domingo. It opens up prospects both in terms of their governance and management and also in terms of academic research.

Regarding governance first, explicitly considering behaviors of individual actors leads to clearly reintroducing the role of emotions in decision-making.

Contrary to a widespread view, family shareholders, depending on their nature and their degree of *affectio societatis*, make a trade-off between financial profitability and emotional profitability, between financial value and emotional value. They are torn between a long-term vision and their short-term selfish instinct. The resulting conflict is similar to the agency conflict that arises between the emotional and rational aspects of an individual's personality. This segmentation of family shareholding around the trade-off financial profitability versus emotional profitability leads to a duality of the cost of capital in family businesses and therefore of a company's value.

Thus, this segmentation gives rise to different types of conflicts also between family members. Conflicts can arise between those who benefit from private gains, particularly emotional ones, and others. These conflicts can arise during a transfer, since the former will be more reluctant than the latter to accept it, because they run the risk of losing their management activities and the profits associated with them. The latter may be attracted by the prospects of financial gain on their securities. The example of the sale of Château d'Yquem to the LVMH group in 1996 is particularly significant from this point of view (Hirigoyen 2014).

The typology of family shareholding makes it possible to highlight a duality of a company's value: a subjective value for family shareholding with a profound *affectio societatis* due to the existence of emotional value; an objective value for a passive family shareholder. This duality modifies the prospects of shareholder governance in a value-based management approach in family businesses. In the Petrus case, the homogeneity of shareholding and the transfer of a minority share mean that family shareholders reconcile the dual objectives, which is not the case for Château Angelus, for example.

The inclusion of emotions in decision-making within companies, and particularly family businesses, is new, even though economists are increasingly recognizing their importance in economic behavior. This integration of emotions leads in the context of the transfer decision to favor a counterfactual emotion, regret. Emotional value defined in this chapter as equal to the present value at the emotional rate of return of emotional cash flows constitutes an estimate of emotional regret felt by family shareholders with a high *affectio societatis*.

This approach leads to academic research perspectives in the context of decision-making in behavioral finance. Thus, considering regret in a decision to transfer leads to the integration of the broader concept of aversion to disappointment and therefore the search for a decision leading to its minimization. It can be considered that the sale of a minority share of capital is part of such a perspective. Similarly, taking into account an emotional rate of return opens the reflection on the existence of a psychological cost of equity.

This chapter thus underlines the importance of a subjective approach to the valuations of family businesses and their securities, in particular by raising the issue of the monetary measurement of emotional value. Thus, it greatly contributes to understanding the price formation of family businesses such as the findings in a specific sector of activity like the wine sector. It also strongly reminds us that the valuation of a company is a complex exercise, which can in no way be reduced to the use of a formula, but which requires a thorough understanding of the reality that we seek to evaluate. There is no "true value", but several prices are possible because several points of view coexist.

Petrus continues to develop its long-standing capital by perpetuating its pioneering spirit. Its name, derived from the first Pope, without a *Château*, which has since been copied by others, distinguishes it as a legendary wine. It is also always part of the future with its spiritual dimension. Recently, Thomas Pesquet boarded two cases of the 2000 vintage aboard a space shuttle. While the "classic" Petrus bottle sells for approximately 4,500 euros, the "Space Petrus" was auctioned by Christie's at a million dollars a bottle. A further boost to the value of this mythical property, which had already caught our attention in 2018.

3.8. References

Aurier, P., Evrard, Y., N'goala, G. (2004). Comprendre et mesurer la valeur du point de vue du consommateur. *Recherche et applications en marketing*, 19(3), 1–20.

Berrone, P., Cruz, C., Gomez-Mejia, L.R. (2012). Socioemotional wealth in family firms: Theoretical dimensions, assessment approaches, and agenda for future research. *Family Business Review*, 25(3), 258–279.

Hirigoyen, G. (2014). Valeur et évaluation des entreprises familiales. *Revue française de gestion*, 5(242), 119–134.

Hirigoyen, G. and Weber, P. (2019). Prises de participations minoritaires : entrepreneurs, sachez gérer les risques. *Les Echos* [Online]. Available at: https://www.lesechos.fr/idees-debats/cercle/prises-de-participations-minoritaires-entrepreneurs-sachez-gerer-les-risques-1039205.

Holbrook, M.B. (1994). The nature of customer's value: An axiology of service in consumption experience. In *Service Quality: New Directions in Theory and Practice*, Rust, R.T. and Oliver, R.L. (eds). Sage, Thousand Oaks.

Keynes, J.M. (1936). *Théorie générale de l'emploi, de l'intérêt et de la monnaie*. Payot, Paris.

Morar, D.D. (2013). An overview of the consumer value literature perceived value. In *International Conference: Marketing from Information to Decision*, 169–186.

Pierce, J.L., Kostova, T., Dirks, K.T. (2001). Toward a theory of psychological ownership in organizations. *Academy of Management Review*, 26(2), 298–310.

Samuelson, P.A. (1968). *L'Économique*, 7th edition. Armand Colin, Paris.

Sumantri, Y. and Lau, S.K. (2010). Desired and received customer value models in logistics service outsourcing. In *Proceedings of the 1st International Conference on Business and Economics*. UP Organizer and Publication Co. Ltd., University of Wollongong, Wollongong.

Zellweger, T.M. (2005). Total value-towards the true value of privately held and family firms. In *Proceedings European Academy of Management*, Munich.

4

Evolution of French Wine Companies in the Global Landscape

4.1. Introduction

Is France, the world leader in the wine market, the leading historical exporter by value, a country whose wines have an unparalleled reputation, in the process of falling into the ranks? Worse yet, won't it be overtaken by its main competitors in the coming years? These questions, which may have seemed provocative a few years ago, are now legitimate, as it appears that French wine companies are struggling to adapt to the changes in the international landscape. These mutations are of a triple nature.

First, it is the people who consume wine and the way they consume it that impact French wine sales. The consumer of the 20th century and that of the 21st century are profoundly different. The first consumes a product, while the second wishes to live an experience. Winemakers must adapt to this. Behind these developments, there is also an accelerated globalization of wine consumption affecting countries that are increasingly distant from Europe, the cradle of wine.

Secondly, the globalization of the wine sector is also that of production. Producing countries have multiplied greatly, and some are experiencing or will soon experience strong growth. Firstly, European countries, with Italy in particular gaining importance. Tomorrow, with global warming, it may be England that will become a competitor to France. But competitive pressure is exerted especially far from us, in the so-called New World countries. The increase in exports from countries such as Australia or Chile far exceeds that of European countries in general, and France in particular. The evolution of competition, emergence of new

---

Chapter written by Jean-Marie CARDEBAT.

countries, with modern and effective marketing and commercial methods, clearly weigh on the commercial performance of French winemakers.

Finally, exporting in today's business environment becomes a real fighter's journey. Macroeconomic changes, with the return of inflation and the specter of recession as well as climate shocks or health shocks create an extremely deleterious business climate for winemakers. We only have to look at the international markets between 2019 and 2022 to understand this situation: in 2019, Trump tax; in 2020, Brexit; in 2022, the war in Ukraine and closure of the Russian market, all against the backdrop of the health and economic crisis related to Covid-19, which started in early 2020.

These changes in competition, consumption and the business environment must lead French companies in the wine sector to adapt their strategies to maintain their leadership on a global scale.

The purpose of this chapter is precisely to characterize these changes and deduce strategies to deal with them.

The first point recalls the evolution of international wine trade over the last 25 years and questions the sustainability of France's leadership. The second point describes the change in world wine consumption. The next point traces the changes in the business environment. The conclusion proposes a strategic reflection on the competitiveness of French wines for export.

4.2. France in the world wine trade: a contested place?

Since the 1990s, the wine market has become globalized. In 20 years, world wine exports have almost doubled, from 60 million hectoliters in 2000 to 111.6 million in 2021 (OIV 2022), despite a slowdown in the 2010s (see Figure 4.1). The increase is even more impressive in value with, over the same period, an increase from 14 to 35 billion euros in the total value of world exports.

Beyond these amounts, it is above all the number of countries participating in the exchange that marks this globalization. Before the 1990s, wine was a relatively little traded product on a global scale. Trade was actually concentrated in six countries. Three major exporting countries (France, Italy, Spain) and three major importing countries (Great Britain, Germany, Belgium). As these countries are European, the problems of barriers to trade were very limited. During the years 2000–2010, the United States and China were the major importers. Asia, Russia and America became the major markets, while Britain exited the EU.

Therefore, we understand the changes at work in this market and necessary adaptations. Large exports have replaced exports to neighboring countries that are members of the same political union. The rules are more complex. Cultural differences induce different marketing.

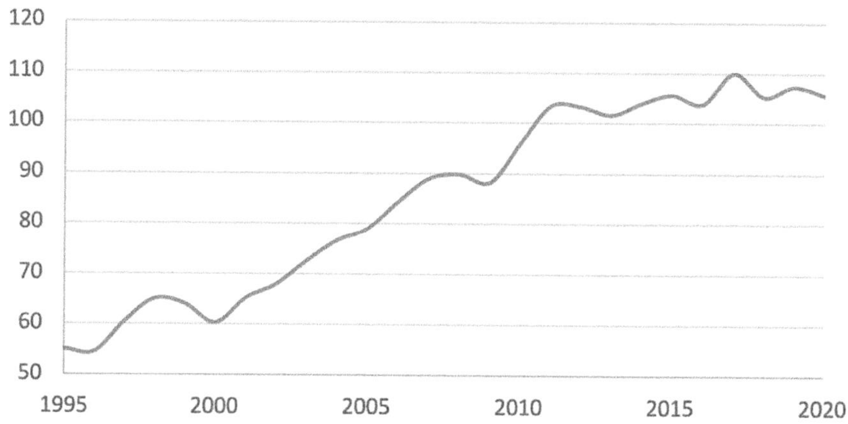

Figure 4.1. *Evolution of world wine trade 1995–2020 in million hectoliters (source: OIV). For a color version of this figure, see www.iste.co.uk/cheriet/wine2.zip*

French producers are facing this new situation in the world wine trade. However, for them, export represents a necessity in the face of a national market in decline since World War II. How have they gone through this phase of market globalization and what is France's place in world wine trade?

4.2.1. *A globalization of the wine market in two or three stages?*

The first wave of globalization of the wine market is historic. It takes us back to the Renaissance and the days when English ships loaded barrels of port and burgundy into their holds as they sailed up colonies to England. This story is well known, and the export market remained essentially European until the 1990s: a game with six countries as mentioned previously.

From there, a new wave of globalization will arise. This is the emergence of the New World, consisting of six countries: United States, Australia, South Africa, Argentina, Chile, and New Zealand. Figure 4.2 shows the evolution of volume exports of the main wine-producing countries between 1995 and 2019. It reveals the impressive progress of these New World countries. With growth rates of three or even four figures, these export rates clearly symbolize this second wave of globalization. Australia and Chile are very serious competitors in the international wine market for the traditional countries of the Old World (France, Italy, Spain).

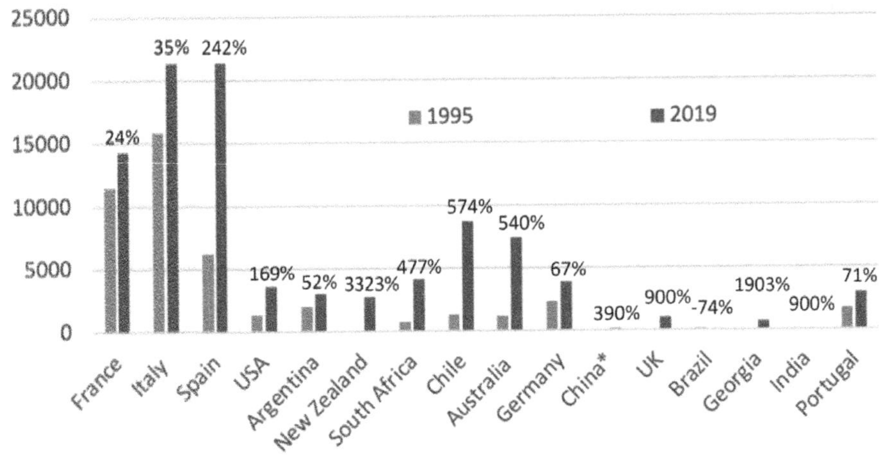

The figures above the bars of the histogram represent percentage increases over the period 1995–2019. *For China, we include wine exports from Macau and Hong Kong.

Figure 4.2. *Wine volumes exported by major exporters in thousands of hectoliters, 1995–2019 (source: OIV). For a color version of this figure, see www.iste.co.uk/cheriet/wine2.zip*

The third wave of globalization remains potential and is slow to materialize. It would come from emerging producer countries, foremost among which is China. We also think of Brazil and India. The production capacity of these countries and the potential size of their domestic markets suggest strong growth in the coming years. However, the statistics of recent years show a stagnation of production in these countries, or even a decline in China, where the fashion phenomenon around wine is reportedly fading. However, climate change could also play a role in the new mapping of international wine production and trade over the next 10–20 years. In particular, Northern Europe could eventually establish itself as a very competitive area.

4.2.2. *France, a leader threatened or overwhelmed?*

However, the trio of traditionally producing countries (France, Italy and Spain) remain the clear leaders in the global wine market. In volume, and even more so in value, they represent approximately 55% (in volume) to more than 60% (in value) of world exports. On the other hand, the hierarchy is more ambiguous among the three countries of this leading trio. This time, it is necessary to differentiate the value of the volumes exported. It should be noted that the "competitiveness" ranking of FranceAgriMer (2022) is a valuable source of information for an in-depth analysis of the competitiveness of the sectors of the main producing countries.

As shown in Figure 4.2, Spain is the leading exporting country by volume. However, it is followed by Italy. But above all, when we consider the value of exports, Spain lags far behind France and Italy. In terms of value, France clearly takes the lead. This country accounts for almost a third of the total value of world exports. According to the OIV (2022), French exports amounted to €11 billion in 2021, compared to €7 billion for Italy and €2.8 billion for Spain, with all other countries being well below €2 billion in exports. What do these figures tell us?

Firstly, Spain does not place a high value on its wines. Spain leads in the export of bulk wines (containers exceeding 10 liters). Bulk wines are by nature undervalued. France stands as the top importer of these Spanish wines. They are then bottled and valued through brands that are often sold for export (Cardebat 2017; Alonso et al. 2019b). These Spanish wines offer exceptional competitiveness, with prices per liter approximately 50 euro cents. The main supplying region is south of Madrid in the Castilla-La Mancha region. Spaniards are allowing a significant amount of value to slip away by exporting these bulk wines. The key for this country is to elevate its position in the market to better capitalize on its production. Efforts are underway, and certain regions are showing promising advancements, but the journey is lengthy and challenging.

Is France, then, the champion when it comes to valuation? Yes and no. Yes, because the international trade figures in terms of value support this claim. Furthermore, the country's image and its most prestigious wine regions strongly advocate for premium valuation. Grand crus are experiencing excellent export performance, with prices showing a consistent upward trend over the past few decades. Nevertheless, the evident success in the high-end market only partially conceals the challenges faced in the entry-level segment. France is facing difficulties in selling its lower-priced wines and effectively enhancing their value through strong brands. In the entry-level segment, the lack of a strong brand and the division of the product into numerous appellations and domain names make the image and information presented to the consumer less clear. This atomicity of the market also prohibits any form of economy of scale and partly justifies a price competitiveness lower than that of its competitors.

Italy stands out as a country that offers an intriguing compromise between the capability to export large volumes (almost on par with Spain) while also adding value (Italian exports in terms of value are 2.5 times greater than those of Spain). In the near future, Italy has the potential to emerge as the primary exporter, both in terms of volume and value. This is particularly significant given that Italy has held the position of the leading producer since the mid-2000s, surpassing France. Finally, the country ranks first in FranceAgriMer's "competitiveness" ranking (2022).

Looking further ahead, the potential challenge for France may lie in a different direction. Indeed, the production trends indicated in Figure 4.3 suggest the possibility of a catch-up, or even a shift in global leadership in favor of the New World. Figure 4.3 shows a very clear long-term trend: the two worlds converge in terms of production. A straightforward extrapolation of the trend indicates that, in the medium to long term, New World production will likely become equal to, and eventually surpass, that of the Old World. Considering the export growth rates of New World countries, it is inevitable that they will emerge as the leaders in the international wine trade. Morrison and Rabellotti (2017) showed that catching up with the New World is inevitable and will inevitably have consequences in terms of leadership. So, there will be a shift towards hegemony in the world wine market.

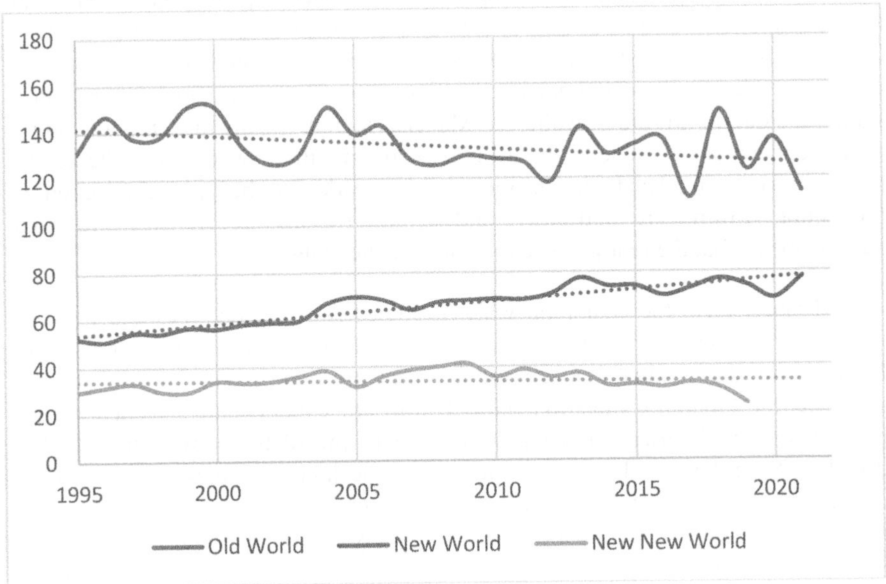

Figure 4.3. *Wine production by major producing regions in million hectoliters, 1995–2021 (source: OIV). For a color version of this figure, see www.iste.co.uk/cheriet/wine2.zip*

The strengths of New World countries are well known (Alonso et al. 2019a; Niklas et al. 2022). Compared to France, their competitive advantage is derived from the larger average size of properties. This larger size therefore allows for better exploitation of economies of scale. Beyond that, their business models, which are more oriented downstream in the value chain and therefore towards marketing, give them a much more marked sales culture. The ability to create successful brands seems much more developed, compared to the list of major global brands. American, Australian and Chilean brands are among the largest. European brands are absent. However, in the international market, brands are an extremely effective means of development.

The fact that France is not included in the ranking of major global brands raises questions about the country's capability to establish a competitive offering in the entry-level and core range segments. This also prompts questions about the country's potential to sustain its leadership (in terms of value) in the global wine trade over the long term. This situation is a reflection of an ever more upscale specialization. Admittedly, this positioning creates value overall, but it cannot allow all vineyards to survive. Luxury is by nature exclusive. For vineyards experiencing a decline in price competitiveness due to the decline of vineyards and rising costs, but lacking strong differentiating factors or non-price competitive elements, the export situation, much like in the French market, becomes challenging in the face of a significant increase in competitive intensity.

Bordeaux, Champagne and Burgundy are the leading high-end vineyards. However, it is not always easy to value wines well within each of these three regions. This is especially true in Bordeaux, which is a very vast vineyard and one of the largest in the world, with a great diversity of economic situations among the more than 5,000 producers. Likewise, prestigious and highly valued estates can be found in most other French wine-producing regions, notably in Côtes du Rhône and Provence with their rosé wines, as well as in the Loire, where the wines align with the current preference for lighter styles.

The evolution of French commercial performance encompasses a diverse and, at times, concerning reality. Without being overwhelmed, France must rise to the entry-level and core range challenges. But this requires a more detailed understanding of consumer expectations, which has changed significantly in recent years.

4.3. Metamorphosis of the global consumer

The globalization of the wine market is also the tale of how this beverage became popularized worldwide. Wine became a global product during the 1990s and 2000s. The consumption boom in Asia, and more particularly in China since the beginning of the 2000s, marked a turning point in the international wine trade.

4.3.1. *Shift in global wine consumption out of Europe*

In less than 10 years, China has established itself as one of the main outlets for wine exporters, especially red wine. This market has become essential for Bordeaux exporters, who are leaders in this market, and for Australian producers. The downturn in the Chinese market since 2018, as shown in Figure 4.4, consequently poses a threat to numerous vineyards, particularly those that had primarily directed their exports towards this destination.

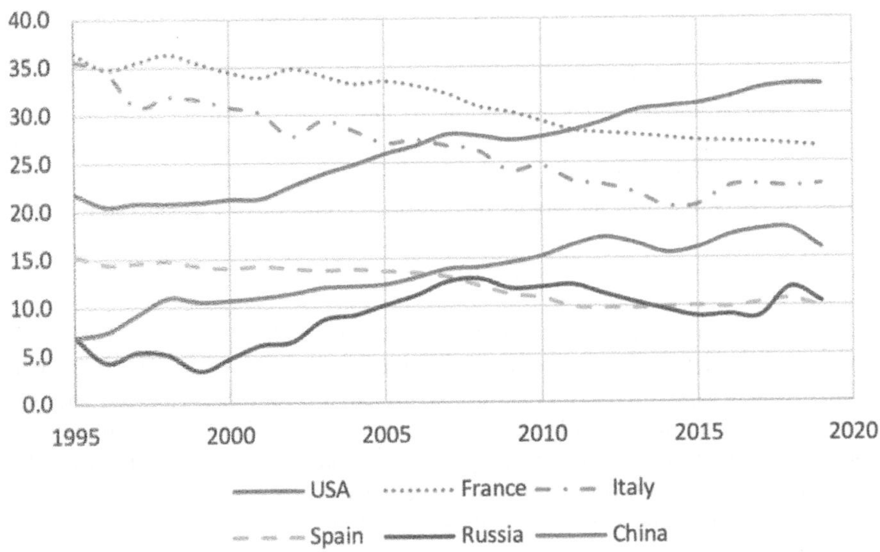

Figure 4.4. *Evolution of wine consumption in millions of hectoliters for a selection of countries (source: OIV). For a color version of this figure, see www.iste.co.uk/cheriet/wine2.zip*

While the emergence of the Chinese market symbolizes the globalization of the wine sector, it is undoubtedly the increase in American consumption that is the trigger. Since the early 2010s, the United States has established itself as the leading consumer market on a global scale, ahead of the large historical consumers such as France or Italy. Figure 4.4 clearly shows the shift in national consumption trends. Since 2006, the dynamic is now outside Europe (see Figure 4.5). Apart from the differences in scales (with the left representing the rest of the world and the right representing total world consumption), the curves for "rest of the world" and "world" indeed exhibit the same dynamics and movement.

Should market dynamism and growth be sought outside Europe for exporters? Partly yes, but the fall of the Chinese market since 2018 and the revival of European markets since 2021 put this conclusion into perspective. Likewise, the sudden shutdown of the Russian market due to European sanctions related to the war in Ukraine serves as a reminder that nothing is guaranteed in the realm of international trade.

The consumption trends shown in Figures 4.4 and 4.5 may not necessarily remain identical in the next decade. French exporters must therefore above all ensure that they diversify their exports so as not to concentrate their efforts on markets that are ultimately volatile.

Conversely, the transformations in consumption over the past 25 years signify a paradigm shift of the wine consumer's sociology. By definition, it is not the same average consumer who buys wine in the 2020s as in the 1990s.

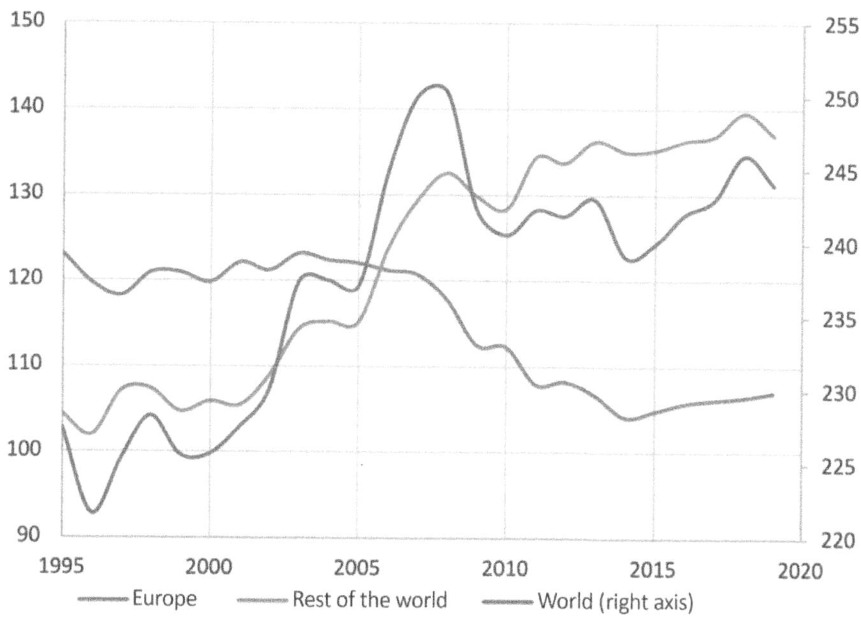

Figure 4.5. *Evolution of world wine consumption and by major regions in millions of hectoliters (source: OIV). For a color version of this figure, see www.iste.co.uk/cheriet/wine2.zip*

4.3.2. *Typical profile of the average consumer in the 20th and 21st centuries*

As consumers have become more international, the average consumer has undergone significant changes. But many other elements have changed beyond nationality. Table 4.1 provides a typical profile of the average wine consumer in the 20th and 21st centuries. While inherently simplifying, this stylistic exercise still sheds light on the ongoing transformations.

Characteristics	20th century	21st century
Distance ratio	Consumption of local wines	Consumption of wines from around the world (with a local bias for producing regions)
Origin	European	Global (mainly American, Chinese, European)
Age	Mostly old	Mixed (largely young)
Gender	Male	Feminization of consumption
Social class	Popular/elite (dichotomous consumption)	Mixed (again with bias towards higher socio-professional categories)
Wine type and distribution	– Basic wines (grand crus for the elite but low volumes), bottles – Hypermarkets and supermarkets (GMS)	– Highly differentiated mid-range wines (variety research) and more whites, rosés and sparkling wines – Packaging and more varied distribution channels
Consumption frequency	– High to daily (with meals) – Functional consumption	– Occasional (party, friends, etc.) – Outside of meals and during meals – Experiential consumption
Degree of information	Either poorly informed or knowledgeable (elite)	Significantly increased thanks to the Internet (smartphone applications, etc.)

Table 4.1. *Main characteristics of the average wine consumer over time (source: adapted from Cardebat (2017, p. 40) and the annual publications of Wine Intelligence)*

As already indicated, it is first of all the origin of wine drinkers that has changed profoundly, with consumption outside Europe having exceeded European consumption. It is worth mentioning that tourism has the effect of inflating the consumption figures for France, Italy and Spain, although quantifying this impact is inherently challenging. These three countries are indeed among the world's top tourist destinations, with many non-European tourists consuming wine and inflating consumption statistics. This phenomenon is certainly far from negligible and accentuates the observation that consumption has moved out of Europe.

Due in part to changes in consumption patterns, there is also a trend towards a more female-dominated consumer base, particularly in English-speaking countries, and a younger consumer base, a trend particularly pronounced in Asia (Cardebat 2017). However, a younger consumer base is not observed in traditional consumption countries (France, Italy, Spain), where the average age of regular consumers tends to increase. Similarly, consumption has become more democratic and tasting is no longer reserved for the elite and the upper class. Tasting clubs have been started everywhere. All of these trends favor a much more varied consumption. The wines consumed belong to the core range category, characterized by a significant level of horizontal differentiation. These conclusions can be found in all reports published regularly by various organizations observing the wine market, in particular the reports of Wine Intelligence-IWSR, Sopexa's Wine Trade Monitor and FranceAgriMer's consumption surveys for the French market.

If we follow these different reports, we can summarize the international market as follows:

– Rising trend in the consumption of organic wine, including biodynamic and natural wines. This trend varies from country to country but represents a significant shift. Organic wine still plays a crucial role in product differentiation. However, it is foreseeable that in the near future, it will become an essential rather than a distinguishing factor. French winegrowers must adapt quickly to this changing landscape, as there is a risk of market share erosion, especially in exports.

– Shifts in the types of wines being consumed. In summary, there is a decrease in red wine and "sweet" wines, but an increase in other categories, such as white wines and rosés, particularly in the United States and France. Sparkling wines are also on the rise, with Prosecco rapidly gaining popularity worldwide and Champagne maintaining excellent health. Additionally, this quest for diversity extends to the acceptance of new packaging formats. Bag-in-box (BIB) packaging, for example, has seen substantial global growth, even in traditional wine-producing countries since the health crisis. Cans are experiencing an impressive boom in the United States. Finally, there is diversity in distribution channels as well. Wine merchants regain market share at supermarkets (large- and medium-sized stores) and, above all, e-commerce has spread widely, with a peak during the health crisis.

– The quest for "local" and proximity is another noteworthy trend. Consumers are seeking a more direct connection with producers, even if they are geographically distant. This reflects the idea of a club that we mentioned earlier. Therefore, consumption should carry a certain meaning and embrace shared values with both the producer and other consumers. This phenomenon may explain the emergence of digital communities centered around certain wines, such as natural wines. The marketing implications are profound because what matters less is geographical distance and more so the alignment of values and a sense of belonging to a community of thought.

– "Premiumization" of wine purchasing is driven by the increase in occasional consumption, which has become much more common than regular or daily consumption. As often noted, people are buying less, but they are buying higher-quality or, at the very least, more expensive wines.

In summary, and in a somewhat exaggerated manner, one of the ideal wines for the typical consumer might be a premium organic rosé available for purchase online, recommended by a community of wine enthusiasts. However, beyond this caricature, there are essential elements worth noting. In particular, the last point deserves further attention.

Table 4.1 highlights the significant increase in consumer information. The Internet enables consumers to interact with one another, particularly through smartphone applications that allow you to leave comments and ratings for all of the wines you have tried. Image recognition systems via smartphones make it possible to scan any bottle. With 60 million users in 2022, the Vivino app leads the market in this regard. They have a vast database with tens of thousands of wines that can be accessed at any time. This means that there are tasting notes and comments available for all these wines. The level of consumer information, contributed by peers, in other words, fellow consumers, is unparalleled compared to previous years.

For an economist, this development is pivotal because the wine-buying process is characterized by an information imbalance between producers and consumers regarding the actual quality of the wine. Applications of this kind partially address this issue by providing immediate and free information to potential buyers. This substantially enhances market efficiency and, consequently, expands the market. This is highly positive news for winemakers. This digital revolution has already unfolded in the hospitality and restaurant sector through apps like Booking.com or Trip Advisor. It is now happening in the wine market, where several recent studies demonstrate that the impact of Vivino ratings on wine demand surpasses the influence of ratings from traditional experts, such as magazines and leading connoisseurs, and so on (Oczkowski and Pawsey 2019; Bazen et al. 2022). Vivino, with four out of five stars, becomes the new benchmark in the market. This is at least the threshold indicated by some French exporters to be able to enter the American market more easily (Bazen et al. 2022).

All of these changes are shaping current and, even more so, future trade. French exporters need to consider these underlying market trends. However, in our country, vineyards are primarily focused on the upstream stages of the value chain (Alonso et al. 2019a), which involve grape and wine production. Often, sales are handled by intermediaries (traders, etc.). Not being directly engaged with the market can limit their awareness of major market trends, which proximity to the end of the value chain (importers, retailers and end customers) would otherwise provide. New World producers are more closely connected with these stakeholders, as they typically manage sales and exports themselves.

One of the key challenges for French producers' export performance lies in their adaptability to the global market. This presents a delicate task, especially when upholding the concept of terroir, where the product is shaped by nature and cannot easily follow consumer trends. Terroir, represented by the system of Appellations d'Origine Protégée (AOP), inherently guarantees uniqueness and serves as a powerful marketing tool. However, the proliferation of AOPs today raises questions about the level of information they offer to the average consumer. Striking a balance between preserving terroir and adjusting to the global market is therefore essential.

The shift to "French wine" is a response occasionally chosen by winemakers in this discussion. The flexibility it offers in adapting to the market and expanding product lines with sparkling wines, rosé or white wines indeed allows them to better meet export demands. However, this approach necessitates having a sufficiently strong brand that can stand on its own without the support of an appellation and is thus recognized by consumers.

However, adapting to consumer expectations does not resolve all of the issues, as the challenges of export can be entirely external to the market itself. The stark reminders of this come from the Covid-19 health crisis and the war in Ukraine. Both have underscored the harsh realities of international business environments and the evolving landscape of consumption and have undergone profound transformations in recent years.

4.4. Disruption of the international business environment

Following a peak in globalization around the turn of the 1990s and 2000s, barriers to trade have continued to proliferate. This phenomenon, increasingly labeled as "de-globalization", affects all sectors and is reflected in international trade figures, which have been decreasing as a percentage of GDP since the mid-2010s (Jean 2019). Despite its robust internationalization dynamics, the wine industry is not immune to this trend.

The current decade may witness a significant decline in large-scale exports. In fact, if we refer back to Figure 4.1, a noticeable deceleration in global wine exports by volume is evident throughout the 2010s. The statistics provide a clear picture: export growth between 1995 and 2010 stood at 75%, but it barely surpassed 2% from 2011 to 2020. This trade slowdown is underway, although the rising prices of wine due to premiumization inflate the value of export statistics. This substantial decrease in exports is primarily attributed to a deteriorating business environment that has hampered the sector's internationalization progress given the existing level of supply.

Several types of barriers affect wine, which in fact appears to be a perfect target for protectionist policies.

4.4.1. *The rise of protectionism: from tariff barriers...*

The context of increasing protectionism is widespread. The trade conflict between China and the United States, which began with the arrival of former US President Donald Trump, has not subsided since his departure. Instead, it appears to be expanding globally, leading to significant tensions in major regions. Why would wine be singled out more than other products in these trade disputes?

Firstly, because wine is considered an agricultural product. Agriculture, globally, is the most heavily protected sector (Anderson 2016). Consequently, wine falls under the umbrella of high customs tariffs that impact all agricultural products. However, it can potentially benefit from preferential trade agreements. For instance, Chile is notably successful in securing favorable terms through bilateral trade agreements, having signed 42 of them. As a result, tariff barriers affect exporting countries differently. Chile and Australia, for example, often enjoy more straightforward access to third-party markets compared to European countries (as we will explore later, such access can also be abruptly restricted[1]).

In particular, European countries that are part of the Airbus consortium were hit hard by the 25% Trump tariff on still wines at the end of 2019. Since 2020, all European Union countries have been affected by Brexit. While a temporary agreement has been reached for wine, the implementation of new customs clearance rules is being established and is proving to be burdensome and costly for European exporters to the United Kingdom. Additionally, since January 2022, China has tightened regulations for exporters. Lastly, European sanctions against Russia, in response to the invasion of Ukraine, effectively bar access to the Russian market. These four markets – the United States, the United Kingdom, China and Russia – are

1 See the World Trade Organization website.

crucial for French exporters and their European counterparts. These protectionist measures, therefore, partially account for the challenges faced by some of these exporters. This is especially true for those with limited resources in terms of time and expertise to navigate these regulatory complexities or those with narrow profit margins that see their price competitiveness erode in foreign markets.

These simmering trade disputes often stem from political or geopolitical tensions. All of the cases described above are driven by this political dimension. However, it is not just European countries that are feeling the impact. Australia, which has experienced heightened political tensions with China, has also faced a significant increase in tariffs on its wines since March 2021. With tariffs rising between 116% and 218% on bottled wines, the Chinese market has effectively closed its doors to Australian exporters. For Australian producers, who relied on China for nearly 40% of their total exports, this was their largest export market. The scale of this shock is immense and underscores the importance of diversifying our export portfolio, as shifts in trade policy can be sudden and far-reaching.

The rise of protectionism is clearly evident and ultimately impacts all exporters, especially when the sector holds significant economic importance for these exporting nations. Targeting countries like France or Australia is significant due to the economic significance of the wine industry in these nations. Wine is also an effective target for several reasons, including its substantial contribution to the trade balance of certain countries. Additionally, the cultural dimension plays a role, particularly for European and French exporters. France is synonymous with wine, and targeting the wine industry is a way to impact an emblematic sector. This is why wine is often singled out, as seen, for example, when France declined to participate in the second Iraq war in the early 2000s. Wine exports then become a highly sensitive trade policy issue, serving as both a numerical and symbolic target of retaliation.

4.4.2. *...to nontariff barriers*

There are also other methods to erect barriers against wine exporters, given that wine is a natural product subject to strict health standards. Sanitary and phytosanitary protectionism, a latent form of protectionism, relies on consumer protection measures to either prohibit or complicate imports through, for instance, chemical tests or specific labeling requirements that increase costs and delays. Substances like sulfites or pesticide residues may indeed become significant challenges for exporters in the years ahead, as their presence in bottles could ultimately lead to new regulations.

Environmental barriers pose another significant challenge. “Green” protectionism, where environmental issues are insincerely used as a pretext to reduce wine imports, is likely to gain momentum. Proving the insincere aspect is very difficult. Like health barriers, these measures fall outside the scope of WTO regulations. Concerns about pesticide use and the carbon footprint associated with wine production and transportation, especially in bottles, raise questions about the environmental impact of this trade. The imposition of a carbon tax in the coming years could potentially become a significant hindrance to bottle exports.

The entire logistics of the industry will need to be reconsidered, from the design of packaging to transportation methods and distribution. Environmental concerns in logistics are not mere excuses for protectionism; they are highly significant and could serve as strong points of differentiation. Companies that effectively manage the environmental impacts of their supply chains could gain a competitive edge over others and tap into markets that prioritize these issues (Northern Europe, Canada, etc.).

Lastly, wine is an alcoholic product, accompanied by a range of negative externalities such as accidents, health issues and violence. Various organizations emphasize these externalities to advocate for stricter sales and consumption regulations. Consequently, this serves as a significant barrier to entry for markets that impose heavy taxation and regulation on the wine industry. In any case, it provides an additional argument for those seeking to decrease wine imports.

As it can be understood, wine presents itself as a multi-faceted target for protectionist trade policies. The business landscape is undeniably becoming more challenging. This challenge extends beyond the wine sector, but it is particularly pronounced within it. Political developments in numerous countries (the rise of populism, etc.) and geopolitical shifts do not indicate any forthcoming reversal of these trends. The 2020s, perhaps even more so than the 2010s, will prove to be a challenging period for exporters, especially those engaged in large-scale exports (outside Europe for French exporters). Therefore, the favorable export figures for 2021 should not lead to excessive optimism within the industry.

4.4.3. *Distance barriers*

Without delving into protectionism and thus deliberate policies aimed at reducing imports through customs tariffs or various regulations, there are inherent barriers to international trade. Economists encapsulate these barriers under the concept of “distance cost”. In reality, this notion of distance encompasses several dimensions.

This is primarily a physical issue: the significant increase in transport costs since the Covid-19 crisis and the invasion of Ukraine serve as a reminder that distance comes at a price. The era of nearly zero marginal transportation costs during the peak of globalization seems like a distant memory (Friedman 2005). For high-end wines with a high unit price, this cost remains relatively low. However, it becomes a significant factor for entry-level wines, where the price increase due to transportation may lead consumers to choose a different wine. In fact, for undifferentiated wines, where price is the primary competitive factor, a few cents can make all the difference in the final consumer's choice or, upstream, in the retailer's decision on which wines to stock. Once again, the business environment is deteriorating because, beyond transport costs, the entire logistics chain is becoming more complex. Delays are increasing, with ports in China operating at reduced capacity due to the zero Covid-19 policy and lockdowns in certain port cities, and there are issues with the availability of containers.

The distance between populations is not just physical, it is also cultural. However, behind culture, there are numerous variables, including linguistic and religious differences, as well as biological factors that influence tastes. These cultural differences, in particular, can impact the level of trust between populations and, consequently, the extent of trade they are willing to engage in (Felbermayr and Toubal 2010). Additionally, there is a biological distance that evolves over time, shaping differences in tastes. This distance plays a significant role in explaining the export patterns of French wines. The greater the distance, the greater the trade reduction (Bargain et al. 2022). The rapid growth of the Chinese market in the early 2000s and 2010s could be viewed as a trend in this country. The abrupt market reversal in 2018 could be seen as a return to normalcy, reflecting this distance between populations.

We can observe the presence of numerous trade barriers, and virtually all of them are on the rise. These barriers will have an even greater impact on wine exports in future than they did in the past. Adapting to these challenges is not easy for French exporters, but strategies are available.

4.5. Conclusion

To maintain competitiveness on the international stage, the entire industry acknowledges the importance of creating strong brands, both in the entry-level and mid-range segments (FranceAgriMer 2021). These brands should have the capacity to compete with those from the New World, which currently dominate the global market. What is required are robust and diverse brands that can cater to various market segments and appeal to a wide range of consumers. Focusing solely on premium or luxury products cannot be the sole strategy, as it is inherently exclusive and effective for only a minority of producers.

Establishing strong brands in entry-level and mid-range segments will necessitate substantial coordination efforts, possibly even significant industry consolidation. This consolidation is likely driven by economic and demographic forces, especially considering the numerous farm transitions expected in the 2020s. The ongoing natural consolidation (as evidenced by data from recent agricultural censuses, particularly those from 2010 and 2020) may present an opportunity in this regard.

While brand building is crucial, it cannot be achieved without innovation. This innovation needs to be versatile and extend beyond just digital and precision agriculture. It encompasses managerial, organizational and societal innovations as well. These aspects are intertwined, and digital technology, in particular, offers opportunities for innovative sales approaches. It enables the creation and engagement of virtual communities to foster a club-like atmosphere and to build customer loyalty, whether they are geographically close or distant. *Blockchain* and NFTs (Non-Fungible Tokens) are becoming essential tools to provide consumers with export traceability, shared security and direct connections with producers. NFTs also offer the potential to significantly enhance the wine consumption experience by incorporating additional services and/or artistic elements. One ongoing transformation is the shift towards "direct-to-consumer", a novel way of communicating and selling, even in the export market, thanks to technology.

Societal innovation also encompasses ecological transition and, more broadly, involves engaging with stakeholders and taking on environmental responsibilities by creating ecosystem services. This extends well beyond the scope of organic practices. An ambitious socio-environmental policy represents a significant shift in approach, but it will undoubtedly be a crucial factor for success in the future. An increasing number of farms are embracing this kind of thinking, which imbues their activities with a sense of purpose beyond wine production.

Lastly, innovation must also encompass organizational aspects within the sector. A thorough examination is needed to simplify offerings in the entry-level segment. While the complexity of AOPs, climates and classifications may enhance performance beyond a certain range, it raises questions about its relevance at the entry level. Exploring options such as revising PDO rules to make them less restrictive (e.g. in terms of climate reserve systems), streamlining market operations more efficiently (e.g. creating a futures market for bulk wines), moving away from mandatory vintage labeling (as seen in Champagne), etc., can help to enhance competitiveness and encourage brand development.

And if we can no longer export as much as before, it is also crucial to consider how to conduct business abroad. The classic strategy for bypassing trade barriers involves foreign investment. If wine cannot be exported, then it must be produced locally. Historically, many estates have pursued this approach in California, Latin America and more recently in China. However, this demands significant financial capacity and is not accessible to everyone. This brings us back to the issue of the fragmentation of farms and their limited financial resources.

Certainly, there is no miracle strategy. In the short term, diversifying exports is the most urgent task, which also incurs costs. Furthermore, a range of approaches must be initiated. Collective action and organization will be crucial to retain international leadership. This leadership will inevitably require a reconquest of the entry-level and mid-range segments, which are slipping away from France in the global market and could ultimately jeopardize its leadership position.

4.6. References

Alonso, A., Cardebat, J.M., Corsi, A. (2019a). *The Palgrave Handbook of Wine Industry Economics*. Palgrave Macmillan, London.

Alonso, A., Cardebat, J.M., Jiao, L. (2019b). The French wine industry. In *The Palgrave Handbook of Wine Industry Economics*, Alonso, A., Cardebat, J.M., Corsi, A. (eds). Palgrave Macmillan, London.

Anderson, K. (2016). *Agricultural Trade, Policy Reforms, and Global Food Security*. Palgrave Macmillan, London.

Bargain, O., Cardebat, J.M., Chiappini, R. (2022). Trade uncorked: Genetic distance and tasterelated barriers in wine trade. *American Journal of Agricultural Economics*, 105(2), 674–708. doi: 10.1111/ajae.12335.

Bazen, S., Cardebat, J.M., Dubois, M. (2022). The role of customer and expert ratings in a hedonic analysis of french red wine prices. Bordeaux Economics Working Papers 2022–17, Bordeaux School of Economics (BSE).

British Medical Association (2014). Have applications to study medicine fallen? [Online]. Available at: http://bma.org.uk/news-views-analysis/news/2014/october/have-applications-to-study-medicine-fallen [Accessed 27 October 2014].

Cardebat, J.M. (2017). *Économie du vin*. La Découverte, Paris.

Cheriet, F. and Maurel, C. (2020). Être petit et réussir à l'export : étude de dix cas d'entreprises vitivinicoles françaises. *Management international*, 24(6), 114–126.

Felbermayr, G.J. and Toubal, F. (2010). Cultural proximity and trade. *European Economic Review*, 54(2), 279–293.

FranceAgriMer (2021). Compétitivité de la filière française vin. Report, La compétitivité des filières agricoles et agroalimentaires, Paris.

FranceAgriMer (2022). Synthèse des facteurs de compétitivité sur le marché mondial du vin. Report, Paris.

Freidman, T. (2005). *The World is Flat: A Brief History of the Twenty-first Century*. Farrar, Straus and Giroux, New York.

Jean, S. (2019). Le commerce international en crise systémique. In *L'économie mondiale 2020*. La Découverte, Paris, 25–42.

Maurel, C. (2009). Determinants of export performance in French wine SMEs. *International Journal of Wine Business Research*, 21(2), 118–142.

Morrison, A. and Rabellotti, R. (2017). Gradual catch up and enduring leadership in the global wine industry. *Research Policy*, 46(2), 417–430.

Niklas, B., Cardebat, J.M., Back, R.M., Gaeta, D., Pinilla, V., Rebelo, J., Schamel, G. (2022). Wine industry perceptions and reactions to the COVID-19 crisis in the Old and New Worlds: Do business models make a difference? *Agribusiness*, 38(4), 810–831. doi: 10.1002/agr.21748.

Oczkowski, E. and Pawsey, N. (2019). Community and expert wine ratings and prices. *Economic Papers: A Journal of Applied Economics and Policy*, 38(1), 27–40.

OIV (2022). Note de conjoncture vitivinicole mondiale 2021. Report.

PART 2

Marketing

5

Storytelling and Narration of Authenticity: The Case of New Winemakers in Saint-Chinian

5.1. Introduction

> It all started in 1910 with Félicien and Fernande who worked 6 hectares, produced their own red wines and became merchants. They passed on their passion to their daughter: Jeannine, who carried with her husband Michel. Together, they undertook the restructuring of their vineyard, which changed to AOP Saint-Chinian in 1982. At that time, Michel worked his vines with a horse.

This is how the Carrière Audier estate presents its history on its website, using a rich collection of photos[1]. Since the 2010s, wine producers have been increasingly opening up, whether it is through their website, social media or during wine tourism visits. They share their personal stories, insights into their profession and even showcase themselves at various events. The aim is to evoke emotions, elicit a lasting response from the audience and establish a connection.

Telling stories is an integral part of daily individual and organizational practices, and it represents a preferred mode of communication. The content of these stories is directly tied to the target audience, so they are often adapted to fulfill specific objectives. Through their communicative power, narratives play a vital role in shaping the identity of the subjects they depict. The act of sharing stories, repeating them and illustrating them are all factors that help to solidify this identity within the

---

Chapter written by Sarah MUSSOL and Yosr BEN TAHAR.

1 See: https://domaine-carriereaudier.fr/notre-histoire/.

audience. This becomes even more evident when examining brand narratives, which are central components of brand-building efforts.

Hirschman (2010) demonstrated that stories are the key to humanizing brands, making them relatable and enabling various stakeholders to identify with them. Narratives serve as a foundation for brand legitimacy, whether it is for a brand, an object or a group. They also play a crucial role in identity construction by highlighting both similarities, group affiliation and the unique attributes of brands. Legitimacy, in this context, is what allows for the identification, classification and categorization of an object or organization, thus providing it with an identity, a sense of belonging and associated attributes.

For instance, in the case of businesses, crafting storytelling imparts purpose to their existence and endeavors. It establishes the context for past and future actions. Within this context, the creation and sharing of stories play a crucial role in shaping the identity traits of an organization, an object or a brand.

A story is a holistic image, comprising facts, emotions and interpretations. These components collectively argue that storytelling serves to persuade a particular reality in which the audience engages and identifies. It functions as a discursive tool for constructing reality, identity and a connection with the audience. Moreover, it serves as a means of communication and exchange, disseminating information that shapes a particular perception of reality and conveys an identity.

When delving into the narratives propagated by organizations, and more broadly into agri-food storytelling, these stories often evoke nostalgia for a bygone era, contrasting it with the present-day modernity. Within the myths associated with this nostalgia, the concept of authenticity emerges, featuring minimally processed products, a natural essence, deep roots in a specific region and terroir, crafted by skilled artisans, renowned for their expertise, and celebrated for their flavor and health benefits. These myths used in storytelling largely stem from marketing strategies rather than the true authenticity of the product or brand. In fact, the notion of authenticity is frequently seen as at odds with marketing, the latter often perceived as an artificializing force that compromises authenticity.

In the unique context of wine and winegrowers who directly sell their products, the issue of marketing their offerings arises. The limited resources available to these organizations can genuinely impede their growth. Leaders, who are primarily winemakers focused on producing wine through the mastery of agricultural and agri-food practices, often tend to neglect marketing tools and strategies. Consequently, the stories shared by winemakers with customers (both individuals and professionals) can become their primary marketing message. This prompts the question of the extent to which storytelling contributes to the cultivation of authenticity.

5.2. Different uses of storytelling

5.2.1. *Narrative structure and narrative engagement*

By infusing emotions, stories serve as a vehicle for imbuing information with meaning (Gabriel 2000). They serve as the preferred medium for generating the essential meaning required for communication among individuals (Boje 1991), "the primary form by which human experience gains its full meaning" (Polkinghorne 1988), whether they are rather real or completely invented (Gioia and Chittipeddi 1991; Weick et al. 2005).

Storytelling is characterized by an actantial scheme (Greimas 1966) that allows for the construction of stories in which an identified hero embarks on a quest with helpers and opponents. In the multitude of stories created by individuals and organizations, certain elements appear to be crucial for successful storytelling in achieving its intended goals: raising awareness, convincing and gaining the support of the target audience. Woodside et al. (2008) propose that successful storytelling exhibits a set of characteristics, including:

– a protagonist taking action to accomplish an objective;

– a story that provides insights into the conscious and/or unconscious thoughts of the protagonist and other actors;

– changes in the protagonist's life;

– events involving the protagonist;

– a provocative event (a crisis or turning point) involving the protagonist where the beginning and resolution are described;

– clear stories.

This specific narrative structure ensures the audience is transported into a particular universe, decoded through mental imagery, thus leading to persuasion. The challenge lies in leveraging story components to captivate consumers, as "stories and their ability to engage the audience constitute the central aspect of the consumer experience" (van Laer et al. 2013, p. 798). The transported audience responds emotionally, experientially and non-cognitively to the story they hear, see or read. The emotion generated forms the core of the connection between the consumer and the consumer object, strengthening this bond (Malita and Martin 2010; Mora and Livat 2013). Stories enable the creation, reinforcement and long-term maintenance of the relationship with the consumer through a clearly defined and strong identity (Holt 2006). The concept of identification relates to the proximity consumers seek between the narrative and their own beliefs and aspirations. As illustrated by the theory of narrative engagement, when individuals become absorbed in a story, "they can demonstrate the story's effects on their beliefs in the real world" (Green and Brock 2000, p. 701).

Narrative engagement places a strong emphasis on the role of protagonists, often the creator or founder, in fostering an emotional connection and attachment between the audience and the story being told. Characters, metaphors and myths constitute the elements of stories that enable the audience to identify with the narrative, form a bond with the characters and evoke emotions (Bonarou 2016; Holt 2003). The entity producing and disseminating the story plays a pivotal role in how the audience receives, analyzes and interprets the narrative, as well as the extent to which this story will have a lasting impact on them.

5.2.2. *Building a narrative identity*

Stories are an integral part of human nature (Fisher 1984) as they serve as a means to acquire legitimacy and construct identity (Edson Escalas 1998). The impulse to share stories is closely tied to the fundamental human need for existence and self-legitimization (Boyce 1996). Legitimation is a process through which both individuals and organizations communicate to their audience the rationale behind their existence, enabling them to align with a specific social group. Through storytelling, they convey concepts, visions, values and institutionalized norms. Legitimacy is conferred by the audience, which assesses the characteristics of the entity and evaluates the extent to which they enable identification within an existing category, as well as differentiation from other entities. Thus, the core of legitimacy construction lies in the notion of identity, a concept that stories help to convey.

Fog et al. (2004) argue that stories imbue people's lives with profound meaning, thus explaining their inherent need to construct, narrate and believe in the stories they encounter. Identity formation can draw upon a diverse array of narratives (Kuhn 2006). Consequently, storytelling also serves as a tool for constructing reality and shaping behaviors. Effective storytelling, whether done consciously or unconsciously, allows strong myths to resonate with the audience, particularly consumers, elevating the brand to an iconic status (Holt 2003). In this context, myths are presented as uncomplicated tales featuring captivating characters and plots that strike a chord. As Holt (2003, p. 44) suggests, "Myths offer guiding ideals and help address life's most intricate questions. Icons are encapsulated myths. They wield significant influence by delivering myths to us in tangible forms, making them more accessible". Myths echo in the collective imagination and, as a result, enjoy broad acceptance, becoming an integral part of shared reality. These factors collectively render myths accessible and relatable, creating stories that are readily embraced and remembered.

Martens et al. (2007) argued that stories are an essential part of the entrepreneurial toolkit, and successful business creators are often those who can craft a compelling narrative. The stories woven by these entrepreneurs result from conscious choices to incorporate real or imaginative elements and structure them within a narrative framework that includes specific settings, timelines and objectives. These choices are made with the explicit goal of communicating a distinct identity and garnering legitimacy from their audience. Authenticity and the elements of authenticity interwoven into the narrative serve as the foundation for grounding this identity and establishing credibility. In artisanal endeavors, the craftsman invariably assumes the role of the central character in their story, embodying both the brand and the product. Their narrative directly relates to their personal journey and strives to demonstrate the exceptional nature of their venture. Being the primary actors in their stories, winemakers cultivate their own identity and that of their wine, an identity intrinsically intertwined with their life paths and choices (Lapsley and Moulton 2013). Through their storytelling, winemakers reveal themselves to their audience, aiming to evoke both emotion and a sense of proximity. Concurrently, by thoughtfully selecting the elements to highlight, they construct the identity of their product.

The wine industry, especially in the realm of wine production and distribution, possesses unique characteristics that make it a rich arena for exploration and experimentation. The key attributes encompass the following:

– The market is very highly fragmented.

– It is difficult to assess the quality without tasting the product.

– Products are poorly differentiated from a consumer perspective.

– Wine has a very strong social image.

– Wine knowledge and its consumption are tied to a masculine myth of a specific age group and a relatively affluent socio-professional class.

– It is a product rooted in the terroir, strongly linked to the idea of naturalness and nature.

In a similar regulated sector, Baillergeau and Benavent (2006) demonstrated that brands exhibit little differentiation in terms of codes, symbols and communication products. The market is then considered a shared space, and the product's construction and social positioning are based on alignment with the social representation of the business. Positioning is determined not by differences in perception but by legitimacy. Winemakers must be recognized as members of the winemaking community. They need to embrace the industry's social norms, particularly to be categorized as winemakers within this market segment. Stories enable winemakers to elucidate the reasons behind their entry into the profession and their motivations for producing wine.

5.2.3. *Different forms of authenticity in stories*

Numerous studies suggest the crucial role of the audience's perception of authenticity in the success of certain alcoholic beverages, such as champagne (Beverland 2006; Miller and Chadee 2008; Vallaster and Kraus 2011) or craft beers (Thurnell-Read 2019). These products convey an image deeply rooted in the producer's expertise, the terroir and the quality of the ingredients. Many vineyards, breweries and distilleries offer immersive experiences to customers, who come to explore the product's history and its maker. Organized tours and tastings serve as occasions for direct interaction between the location, the products and the stories, allowing customers to fully immerse themselves and, most importantly, be transported. This is where the story comes to life and gains even greater significance (Ocejo 2017; Smith Maguire et al. 2017).

In a study examining the narratives shared by craft beer producers, Thurnell-Read (2019) emphasized that the communication implemented by producers through their website and the stories they employ to introduce themselves serves the purpose of conveying the authenticity of the product and the production process. These stories are designed to assert the authenticity of products and enable their positioning in the market, particularly with added value.

Authenticity is regarded as the outcome of a rather artisanal achievement, not innate, which contrasts with mass consumption. Artisanal products convey authenticity in the face of an inauthentic world, offering the opportunity to involve the consumer in the production process (Ocejo 2017). The concept of authenticity is deeply rooted in these genuine products where the process is upheld, and the value is crafted with a mastered love and expertise. According to the findings of Thurnell-Read (2019), producers incorporate six facets of authenticity into their narratives:

– procedural authenticity: emphasizing production and manufacturing procedures and processes showcases the mastery of craftsmanship, a distinguishing factor for producers;

– material authenticity: involves an emphasis on the quality and value of ingredients used in production and on the product itself; the selection of ingredients resonates strongly with this type of authenticity;

– geographical authenticity: localization, local geography and the symbolism of the place are integral to the product's identity, making it unique;

– temporal authenticity: refers to tradition, time in a broad sense or specific time periods. This authenticity aligns with local heritage and tradition;

– oppositional authenticity: by contrasting itself with the perceived inauthentic other (e.g. multinational corporations and large companies) in order to enhance its own authentic value;

– biographical authenticity: this aspect centers on the manufacturer or creator of the product, often describing the product as an expression of the manufacturer's taste and character, rather than solely focusing on the product itself.

The concept of authenticity remains flexible, and as Beverland (2005) highlighted, authenticity is often arbitrarily determined and associated with notions of purity, naturalness, reality, tradition and preserved heritage. It is attributed when certain intrinsic and extrinsic characteristics are present. Authenticity is closely tied to perception and, consequently, evaluation based on standards and social codes associated with the product, making it a subjective concept (Postrel 2003). Furthermore, consumer characteristics play a role in influencing the perception of authenticity. According to Beverland (2005), authenticity is the outcome of the interplay between intrinsic product-specific elements like the product itself, the production process, and/or the connection to place and history, and subjective elements created by stakeholders, such as myths, status and image.

Legacy plays a significant role in shaping consumers' perception of wine authenticity. For instance, showcasing the cellar where the wine was aged stands as one of the key elements in brand strategies associated with authenticity (Beverland 2005). In the wine industry, authenticity often becomes a focal point in winemakers' visibility strategies. The legacy of a particular tradition serves as a foundation, coupled with the utilization of space (place, geography) and time (temporality, process). When comparing Old World wines to New World wines, the former tend to have an advantage, drawing upon their heritage, familial legacy and historical ties. This perceived advantage, signaling quality, translates into better performance, particularly in terms of sales and valuation (Peterson 2005). Conversely, for New World wines, Frost et al. (2020) demonstrate that integrating heritage elements, especially familial, into the storytelling process, along with authenticity components, becomes essential in gaining a competitive edge.

While previous studies have delved into the narrative practices of luxury wines from established houses or wineries with a certain heritage, we are exploring the strategies employed by new winemakers. Indeed, the issues of identity and legitimacy do not inherently stem from inheritance. The preservation of tradition and the storytelling of authenticity among these emerging winemakers are intriguing because:

– The story to be told is inevitably different due to the absence of heritage, reflecting the endeavor to construct a unique winemaker identity, as well as the quest for legitimacy with the audience.

– Creating authenticity involves dealing with paradoxes (Beverland 2005).

– The prominent family legacy in the industry, which serves as a source of legitimacy and authenticity, is missing in this case. The elements of authenticity mobilized are other than those related to inheritance and family history.

5.3. The Saint-Chinian case

5.3.1. *Introducing the case: narratives from emerging winemakers*

The Saint-Chinian AOP comprises 300 winemakers, each with their unique profiles. In 2020, with the assistance of an agency, it established an identity based on the combination of schist and limestone terroirs. This region also attracts newcomers to establish themselves as winemakers due to the relatively affordable land prices. It is of interest to explore how these emerging winemakers integrate into this appellation.

An emerging winemaker is defined as a wine producer who chooses to establish themselves in the industry without a direct family heritage connection to winemaking. They are essentially novice entrepreneurs in the wine business. Emerging winemakers may, nonetheless, have a distant family background related to wine or have prior experience in a related field.

To analyze the storytelling of emerging winemakers and delve into the mobilized authenticity, we have chosen emerging winemakers from the Saint-Chinian winegrower list who meet the following criteria: newcomers to wine production with no direct family ties to the industry. Information regarding their establishment conditions is often presented in various communication channels, particularly on their websites. Our selection process, based on the information available on the winemakers' websites, excluded potential emerging winemakers who either lack a website or do not prominently feature it in their communication materials. We identified a total of 14 emerging winemakers, 11 of which having established their presence in the 2000s. The content from the "our story" or similar sections of the websites within our study sample was gathered and underwent a dual thematic coding process, resulting in the emergence of three key themes: the vineyard, the winemaker and the wine.

The analysis underscores the richness of their narratives and the multidimensional, versatile nature of authenticity. To assimilate into a highly regulated sector, access a network and acquire the necessary resources for their activities, these winemakers evoke a love for the terroir that they share with their fellow winemakers. Consequently, if we apply Thurnell-Read's taxonomy (2019), the primary form of authenticity that emerges from their narratives is geographical authenticity. This plays a crucial role in their identification and the establishment of legitimacy as Saint-Chinian winemakers. Their deep attachment to the terroir and vineyard also explains their involvement in initiatives such as creating collective brands or participating in promotional efforts, such as those promoting the Saint-Chinian vineyard, rather than solely focusing on developing their individual wine brands.

Their identity as winemakers is also closely associated with the utilization of wine production techniques and the distinctive nature of their wine, as well as their commitment to land stewardship. In doing so, they emphasize an authenticity that is both procedural, rooted in the observance of age-old methods, and oppositional, running counter to the prevailing trends in the industry (Thurnell-Read 2019). This can also be viewed as a marketing strategy aimed at establishing a foothold in the market.

Given their lack of a heritage to anchor their narratives, some emerging winemakers turn to the concepts of family and life projects. Biographical authenticity, as outlined by Thurnell-Read (2019), centers around the narrative of their life journey. By recounting their stories as emotionally charged adventures, in which their roles as heroes shift, they also respond to a market demand where various audiences, especially their consumers, eagerly anticipate genuine accounts of producers' lives. In doing so, they affirm the notion that, at times, consumers value a compelling story more than a superior product (Smith Maguire et al. 2017).

5.3.2. *Identifying as a winemaker through geographical authenticity*

There appears to be a unanimous sentiment in the narratives of all winemakers: they operate within an extraordinary vineyard. Significantly, they highlight the vineyard's capacity to yield top-quality products:

> Our 11-hectare vineyard enjoys a prime location on a clay-limestone plateau at an elevation of 300 meters, nestled at the base of the Black Mountains[2] (Domaine de la Femme Allongée).

2 All the quoted passages provided are sourced from the websites of the mentioned estates.

The absence of transmission and inheritance is counterbalanced by a rediscovery of ancestral methods and a reconnection with the land, as if these actions allowed a rekindling of history. Being a winemaker entails working, hands-on, with exceptional land that is treated with respect, and the ability to capture it in a bottle. Authenticity is thus portrayed as a goal of the winemaker's labor:

> Working in the vineyard is the essence of sharing our passion with everyone around us by creating wines which respect the earth, the vines and the grapes whilst finding the correct balance with nature, in order to create an everchanging product (Mas de Cynanque).

> The estate boasts its own bottling equipment, ensuring authenticity (no blending allowed as per the Independent Winemakers‘ charter) and self-sufficiency (Le Mas de Ronnel).

> Our purpose at VitiFarm is to craft an authentic wine that truly embodies its terroir (Domaine Lanye-Barrac).

5.3.3. *Developing procedural authenticity*

The identity of these winemakers is shaped through their choice of techniques. Being a winemaker involves a working method that respects the land and showcases the vineyard:

> They are passionate about traditional winemaking, without high technology or an oenologist, in order to create authentic products with a strong personality that evoke a rustic and magnificent terroir (Domaine de la Femme Allongée).

Many of them choose biodynamics and organic farming, and several emphasize the use of traditional methods against the dominant choices in the industry:

> With a passion for crafting unique wines without concern for trends or marketing, wines that we primarily enjoy, wines meant to bring pleasure… (Domaine Les Eminades).

> Jean-Marie Rimbert, a native of Provence, went against the grain of local customs. A nonconformist, he remains committed to hillside vineyards, championing traditional grape varieties like Carignan, Grenache, and Cinsault (Domaine Rimbert).

Thus, the specific identity of these winemakers will be reflected in the production of unique, original wines, sometimes qualified as authentic:

> The soils teem with life, the vines in perfect harmony with the land, and the wines enriched by these splendid terroirs. Vines influenced by their surroundings yield one-of-a-kind wines that reflect the diverse facets of nature (Domaine Borie La Vitarèle).

> At Mas de Ronnel, this is how we craft wines with a distinctive character, ensuring uniqueness and pleasure (Le Mas de Ronnel).

> The hand picking of the grapes and such attention to detail given to the exact importance of each vine, allows Mas de Cynanque to produce wines made with style and elegance (Mas de Cynanque).

5.3.4. *Narrating biographical authenticity*

Websites, while introducing winemakers, primarily emphasize the connection with the place, often described as a love at first sight:

> When Tine Claeys and Marc Verstraete first saw Château Castigno, they instantly felt that something very special had just begun. After exploring the world, they had discovered one of the best-kept secrets, hidden in the South of France (Château Village Castigno).

> Thanks to a deep connection with the Villespassans terroir, this long-cherished project became a reality in 2006 on a limestone plateau at an altitude of 250 meters with the establishment of Domaine Pech de Lune (Domaine Pech de Lune).

The tale of the discovery and establishment of the estate also unfolds as an adventure story, requiring the winemaker to step into a role:

> It was the start of our adventure, with a little boy of 18 months, called Pascal, and the birth of Rémi who arrived just as we moved into the vineyard (the final stages of the purchase taking place with maternity leave!) (Mas de Cynanque).

As heroes of the story, they are entrusted with a mission – to breathe new life into a place, a vineyard, a terroir. Along their journey, they also encounter adversaries. The challenges of this adventure elevate it, making it all the more noble and deserving of narration:

> Jean-Paul and Albin dedicate themselves to working in harmony with nature, practicing sustainable farming on the 11 hectares of vineyards (Domaine de la Femme Allongée).

> In search of vast landscapes, but strongly discouraged by our banker from pursuing such a venture (we were a bit young to earn his trust), we purchased our initial lands, old vines, and neglected terroirs… (Mas Champart).

> With the financial resources available at that time, the vineyard was revitalized, and 12 hectares were planted (Mas Champart).

These narratives are sometimes characterized by a change in life. This change is often accompanied by a life project that may involve family. Transmission to future generations thus takes the place of the legacy passed down by previous generations:

> Their reunion marked the start of a remarkable new chapter, as if each had lived up to that point solely for the fulfillment of their shared destiny. Since then, they have been inseparable and made the choice to align their lives with the rhythm of the vines they acquired in 2013. They tend to these vines with passion, making biodynamics their guiding principle (Domaine Lanye-Barrac).

> Two other boys (Sylvain born in 2006 and Simon in 2008) made our team complete (Mas de Cynanque).

> The adventure has evolved into a family endeavor, as the children, after completing their studies, pursued additional training to be ready for taking over the Estate when the timing is appropriate (Domaine de Cambis).

Prior experience or the decision to pursue specific training is leveraged to establish a level of legitimacy in this career transition:

> After completing the VITI OENO training in 2004, during which I conducted winemaking internships with fellow winemakers, I made the decision to establish myself (Domaine de Cambis).

Passion remains at the core of their journeys. Many emerging winemakers highlight in their introductions the compelling force of destiny that drove them to pursue their passion, and in some cases, their life's project:

> Nothing in my background led me to embrace the role of a winemaker, but when I acquired the vineyard, it became clear to me,

> given my deep connection to nature, to embark on this adventure (Domaine de Cambis).
>
> Without a family tradition in winemaking, Mélanie, an osteopath in Corsica, and Bernhard, a sommelier in Australia, had nothing predestined them to bring their life project to fruition: that of living and working with passion in nature, and for their backdrop, the vineyard and wine (Domaine Lanye-Barrac).

5.4. Conclusion

Whether the medium for storytelling is oral tradition, illustrated representation or digital platform, the creation and sharing of stories remain a central focus for individuals and organizations. Storytelling narrates the stories of individuals, organizations, and products or brands. It conveys the information that enables their clear identification through specific characteristics.

These stories, entirely crafted by winemakers and emerging winemakers, convey their perspective on the profession, their role and their wine. This act of storytelling focuses on specific aspects of the winemaker's history, the vineyard and practices. It highlights certain elements while blending real and imaginative details. However, its primary purpose is to convey relevant information about the product, its quality, its value and, most importantly, its authenticity. The use of storytelling helps build identity and legitimacy by integrating into the wine sector through discourse. The storytelling of authenticity enables the transmission of emotions and the creation of connections with the audience, especially with (potential) consumers.

However, this form of storytelling does not imply complete transparency for winemakers, as each chooses to shed a certain light on their practices and techniques. This approach initially serves to legitimize themselves and subsequently to set themselves apart. From the producers' perspective, storytelling can be an intriguing strategy to establish a presence in the winemakers' market. Given the substantial number of winemakers sharing their daily narratives, the future of storytelling may involve the development of new narrative forms that preserve a greater sense of mystery. Currently, narrating the various forms of authenticity not only fosters a connection with the audience but, more importantly, legitimizes their standing in the eyes of various stakeholders.

5.5. References

Baillergeau, D. and Benavent, C. (2006). Une approche sociologique du positionnement des marques : représentation sociale et légitimité des marques de surfwear. *Congrès de l'Association Française de Marketing (AFM)*, Nantes.

Beverland, M.B. (2005). Crafting brand authenticity: The case of luxury wines. *Journal of Management Studies*, 42(5), 1003–1029.

Beverland, M.B. (2006). The 'real thing': Branding authenticity in the luxury wine trade. *Journal of Business Research*, 59(2), 251–258.

Boje, D.M. (1991). The storytelling organization: A study of story performance in an office-supply firm. *Administrative Science Quarterly*, 36(1), 106–126.

Bonarou, C. (2016). Visual storytelling & tourism destination branding. In *Conference Proceedings New Media and Entrepreneurship in Tourism*, Panteion University, Athens.

Boyce, M.E. (1996). Organizational story and storytelling: A critical review. *Journal of Organizational Change Management*, 9(5), 5–26.

Edson Escalas, J. (1998). *Advertising Narratives: What are They and How do They Work?* Representing Consumers, Routledge.

Fisher, W.R. (1984). The narrative paradigm: In the beginning. *Journal of Communication*, 34(1), 74–87.

Fog, K., Budtz, C., Yakaboylu, B. (2004). *Storytelling: Branding in Practice*. Springer, Berlin.

Frost, W., Frost, J., Strickland, P., Smith Maguire, J. (2020). Seeking a competitive advantage in wine tourism: Heritage and storytelling at the cellar-door. *International Journal of Hospitality Management*, 87, 102460.

Gabriel, Y. (2000). *Storytelling in Organizations: Facts, Fictions, and Fantasies*. Oxford University Press, Oxford.

Gioia, D.A. and Chittipeddi, K. (1991). Sensemaking and sensegiving in strategic change initiation. *Strategic Management Journal*, 12(6), 433–448.

Green, M.C. and Brock, T.C. (2000). The role of transportation in the persuasiveness of public narratives. *Journal of Personality and Social Psychology*, 79, 701–721.

Greimas, A.J. (1966). *Sémantique structurale : recherche de méthode*. Larousse, Paris.

Hirschman, E.C. (2010). Evolutionary branding. *Psychology & Marketing*, 27(6), 568–583.

Holt, D.B. (2003). What becomes an icon most? *Harvard Business Review*, 81(3), 43–49.

Holt, D.B. (2006). Jack Daniel's America: Iconic brands as ideological parasites and proselytizers. *Journal of Consumer Culture*, 6(3), 355–377.

Kuhn, T. (2006). A 'demented work ethic' and a 'lifestyle firm': Discourse, identity, and workplace time commitments. *Organization Studies*, 27(9), 1339–1358.

van Laer, T., de Ruyter, K., Visconti, L.M., Wetzels, M. (2014). The extended transportation-imagery model: A meta-analysis of the antecedents and consequences of consumers' narrative transportation. *Journal of Consumer Research*, 40(5), 797–817.

Lapsley, J. and Moulton, K. (2013). *Successful Wine Marketing*. Springer, Berlin.

Malita, L. and Martin, C. (2010). Digital Storytelling as web passport to success in the 21st Century. *Procedia – Social and Behavioral Sciences*, 2(2), 3060–3064.

Martens, M.L., Jennings, J.E., Jennings, P.D. (2007). Do the stories they tell get them the money they need? The role of entrepreneurial narratives in resource acquisition. *Academy of Management Journal*, 50(5), 1107–1132.

Miller, K. and Chadee, D. (2008). The relevance and irrelevance of the brand for small and medium wine enterprises. *Small Enterprise Research*, 16(2), 32–44.

Mora, P. and Livat, F. (2013). Does storytelling add value to fine Bordeaux wines? *Wine Economics and Policy*, 2(1), 3–10.

Ocejo, R.E. (2017). *Masters of Craft: Old Jobs in the New Urban Economy*, 1st edition. Princeton University Press, Princeton.

Peterson, R.A. (2005). In search of authenticity. *Journal of Management Studies*, 42(5), 1083–1098.

Polkinghorne, D.E. (1988). *Narrative Knowing and the Human Sciences*. State University of New York Press, New York.

Postrel, V. (2003). *The Substance of Style: How the Rise of Aesthetic Value Remaking Commerce, Culture, & Consciousness*. HarperCollins, New York.

Smith Maguire, J., Watson, D.J., Lang, J.T. (2017). The question of 'alternatives' within food and drink markets and marketing: Introduction to the special issue. *Journal of Marketing Management*, 33(7–8), 495–501.

Thurnell-Read, T. (2019). A thirst for the authentic: Craft drinks producers and the narration of authenticity. *The British Journal of Sociology*, 70(4), 1448–1468.

Vallaster, C. and Kraus, S. (2011). Entrepreneurial branding: Growth and its implications for brand management. *International Journal of Entrepreneurship and Small Business*, 14(3), 369–390.

Weick, K.E., Sutcliffe, K.M., Obstfeld, D. (2005). Organizing and the process of sensemaking. *Organization Science*, 16(4), 409–421.

Woodside, A.G., Sood, S., Miller, K.E. (2008). When consumers and brands talk: Storytelling theory and research in psychology and marketing. *Psychology & Marketing*, 25(2), 97–145.

6

Communication and the Évin Law Relaxation: Exploring the Potential Benefits of Imagining Instructions

6.1. Introduction

Wine consumption is undergoing quantitative and structural changes both nationally and internationally. In France, specifically, advertising for alcoholic products is subject to strict legal regulations aimed at safeguarding public health.

This dual observation naturally leads to the subsequent text discussing, first, the economic and managerial characteristics of the sector, and second, the presence and importance of using mental images in advertising, especially in this particular market.

6.2. Wine market: economic issues, consumption structures and managerial levers

The economic and managerial characteristics of the wine sector indicate a more "challenging" situation than in the past. While marketing practices are increasingly necessary, they must also adapt to a particularly constrained legal environment.

Recently, to illustrate, a debate has arisen within political institutions concerning the regulation of influencer communication on the Internet. The Meta group has even faced legal action due to certain members' posts.

Chapter written by Rémi BRÉHONNET and Jean-François TRINQUECOSTE.

6.2.1. *A market that is now more challenging commercially*

Summary of the conference by Pau Roca, Director General of the International Organization of Vine and Wine[1], April 27, 2022

Global wine exports experienced a 4% volume increase and a 16% rise in value in 2021, resulting in a worldwide wine market totaling 111.6 million hectoliters and 34.3 billion euros. The export index stood at 47%. However, global wine production witnessed a 1% decline overall during the same year. This decline was driven by contrasting situations, including historic drops of 19% in France and 14% in Spain, while the southern hemisphere, benefiting from highly favorable weather conditions, achieved record production levels. Australia (+30%), Chile (+30%) and Argentina (+16%) boosted their production, whereas New Zealand (–19%) suffered setbacks due to frost and hail episodes.

In response to this, global wine consumption made a comeback with a 1% increase in 2021, according to the OIV, following its lowest level since 2002 in 2020. It reached 236 million hectoliters. However, consumption trends varied from one country to another. Consumption remained strong in the United States, while it saw an increase in France, Spain and Brazil. The situation in China was different, as demand had been declining continuously since 2008. The Covid-19 pandemic further accelerated this downward trend, resulting in a 15% decline in 2021 compared to 2020.

Pau Roca also discussed the "bottlenecks" that the industry has been grappling with since the recovery. He outlined the issues related to supply, transportation, energy costs, labor and geopolitical challenges resulting from the conflict in Ukraine. He emphasized the need to "reevaluate the global supply chain". The increasing inflation rate is a major cause for concern, and he suggested that the OIV should closely monitor its impact on wine consumption. Roca mentioned the likelihood of shrinking profit margins, especially affecting budget wines, and a possible decline in demand. He also highlighted the significance of the Russian market as an importer. In 2021, Russia ranked as the 10th largest wine importer by volume, accounting for 2% of global imports. It ranked 8th in terms of import value, with 1.1 billion euros. Italy led the list of exporting countries with a 33% share, followed by France at 19% and Spain at 12%.

Box 6.1. *World Wine Conjuncture 2021 and 2022*

The numbers published by FranceAgriMer[2] or by the CNIV (the trade association for appellation wine marketing boards) indicate a significant and

1 OIV 2021 review of the international wine situation, accessible from: https://www.oiv.int/fr/vie-de-loiv/2021-maximum-historique-en-exportationsnbsp.

2 FranceAgriMer (2016). Les chiffres de la pépinière viticole 2014–2015.

consistent decrease in wine consumption in France. This trend has been fundamental since the 1970s. For instance, French wine consumption has decreased from 100 liters per capita per year in 1975 to approximately 38 liters today. It is evident that over time, the French have become occasional consumers, prioritizing quality over quantity and choosing to consume wine occasionally rather than regularly. As an example, based on figures published by FranceAgriMer from 1980 to 2015, the number of regular consumers has decreased from 63% to 25%, while the number of occasional consumers has increased from 37% to 75%. Additionally, the overall per capita wine consumption has declined from 54.1 liters per capita per year in the 2004–2005 period to 38.1 liters per capita per year in the 2019–2020 period[3].

This shift in quantity is accompanied by changes in behavior. Nowadays, wine consumption is primarily reserved for special occasions with a hedonistic character. These new consumption patterns ultimately create an increasingly challenging situation for the wine industry. While producers and traders are tempted to use commercial strategies, especially in terms of communication, it is essential to acknowledge that these practices must adhere to a specific legal framework at the national level, which includes various prohibitions, such as those related to consumer promotions, as mandated by the Évin law.

6.2.2. *Necessary marketing with constraints: a closer look at the Évin law*

The communication of the alcohol industry, as well as that of wine in particular, has been heavily regulated by French legislation since the enactment of the Évin law in 1991. The then Minister of Health, Claude Évin, initially crafted a law aimed at regulating alcohol advertising and banning it for tobacco. When it comes to alcohol communication, the law mandated that visuals used could only include informational elements such as alcohol content, color or product origin. Consequently, advertisements that depicted images and/or texts suggestive of a positive association between alcohol and pleasure, "glamor" success, performance, sports, sex, opinion leaders and more were effectively prohibited (Gallopel-Morvan et al. 2017). Additionally, Duquesnois and Cheriet (2021) highlighted that the legal framework of the Évin law extends beyond product imagery, imposing constraints on media, semiotics, argument nature and more[4].

3 FranceAgriMer (2020). Direction Marchés, études et prospectives. Les données statistiques de FranceAgriMer.

4 Duquesnois, F. and Cheriet, F. (2021). Loi Évin, marketing du vin et protectionnisme [Online]. Available at: https://openwinelaw.fr/article/view/4747.

In 2015, a unanimous amendment was passed, challenging the strict regulations of the Évin law in its entirety and proposing to distinguish between communication and advertising for alcohol. Under this new legal framework, it is now possible to communicate about the oenological and cultural attributes related to the terroir of wine production. Content, images, representations, descriptions, comments, or references pertaining to a production region, toponymy, reference, geographical indication, terroir, itinerary, production area, expertise, history, or cultural, gastronomic, or landscape heritage associated with an alcoholic beverage holding a quality or origin designation, or protected under Article L. 665-6 of the Rural and Maritime Fisheries Code, are no longer considered advertising or propaganda. The latter indeed specifies that "wine, a product of the vine, winegrowing terroirs, as well as ciders, perries, spirit drinks, and beers stemming from local traditions are part of France's protected cultural, gastronomic, and landscape heritage"[5].

However, this change has loosened the restrictions and allowed for the consideration of new approaches to discussing wine. As an example of these potential new approaches, we can highlight the initiative of the Bordeaux Wine Council (CIVB), which, starting in 2015, launched a new campaign in newspapers, posters and social media platforms worldwide, breaking away from the traditional communication norms of Bordeaux wines. To comply with the new provisions of the Évin law, which prohibit the promotion of alcohol consumption, this campaign uses an illustrative approach that encourages discovery and imagination. It features the iconic 75 cl Bordeaux bottle, accompanied by some revealing "verbatim" statements about the quality of the terroir: "*il y a tant à découvrir… Dans la région de Bordeaux, le sol apporte quelque chose de magique, il offre à nos vins une variété de style que l'on ne voit nulle part ailleurs*" (There is so much to discover… The cool nights and warm days of Bordeaux allow our wines to express their finesse and elegance fully).

Conveying the benefits of the wine-producing region, this form of communication aims to spark and nurture the recipient's imagination, relying on both visual elements and the verbatim content presented in the communication medium. This encouragement to "imagine" is designed to create both personal sensations and mental images. The advertiser has chosen this approach because it is believed to maximize wine communication "within the constraints of the law". Some prior studies provide an academic basis for this industry's optimism. They have demonstrated that explicitly prompting individuals to "imagine" through mental imagery has an impact on their attitude towards the brands/products and

5 Article L. 665-6 of the Code Rural et de la Pêche Maritime. Available at: https://www.legifrance.gouv.fr/codes/article_lc/LEGIARTI000029580897/#:~:text=Le %20vin %2C %20 produit %20de %20la,paysager %20prot %C3 %A9g %C3 %A9 %20de %20la %20France.

advertising in question, in contexts quite similar to this (Babin and Burns 1998; Gavard-Perret and Helme-Guizon 2003).

Nevertheless, as none of these studies have directly examined the impact of these specific techniques in the specialized field of wine communication, this chapter will focus exclusively on investigating the role and impact of "instructions to imagine"[6] on attitudes towards advertising and wine brands.

6.2.3. *Lawful and promising use of mental imagery*

Wine is, by its very nature, an agricultural product and, in a certain way, a cultural "marker", characterized by strong symbolic significance and diverse hedonistic and sensory representations. Its consumption is not merely a mundane act; it often carries elements of imagination and emotion (Château 2009). Furthermore, wine consumption entails a certain inherent complexity, as it engages multiple sensory aspects during its tasting (Croijmans et al. 2019). This complexity adds to the challenge of effectively communicating about this product.

Highlighting the specificities of the terroir is therefore an intriguing communication advantage. Regions and grape varieties indeed possess their distinct identity and character, which can be astutely leveraged to intelligently anchor communication efforts. Several studies have delved into this specific aspect. For instance, Mora and Livat (2013) examined the use of *storytelling* for grand crus and, more precisely, the positioning fostered by the various communication styles prevalent in this domain. Beverland and Luxton (2005) explored the authenticity of wine brand discourse and how they can be developed or reinforced through their communication. Communication strategies within the realm of high-end wines have also been scrutinized, considering the "nature/culture" dichotomy they engage with and drawing inspiration from the phenomenon of artification (Passebois et al. 2015). Boutaud and Dufour (2012), on their part, examined themes related to the "sacred" and the "terroir". Industry professionals discussed these communication aspects during a roundtable on April 14, 2016 (see Box 6.2).

However, a fundamental question persists: what processes occur in the mind of a consumer when exposed to these various messages, and is there indeed a confirmed role for mental images as an underlying mechanism in the information processing process? If such a role is substantiated, what is its significance in terms of attitude towards advertising and the presented wine brand?

---

6 Also referred to in the literature as "visualization induction scripts".

Survey of communication practices implemented by French winegrowers

In the face of intense competition, the Évin law, and the complexity of the systems involved (such as the proposed wine classification, the organization of the industry, the diversity of audiences and markets, etc.), stakeholders in the wine sector have crafted tailored strategies that place the consumer at the heart of the matter. The key attributes emphasized are the intrinsic quality of the wine, the bottle and the label, all of which define the characteristics of the product being marketed. However, the codes, particularly the graphic elements associated with French wine, must be precise and adhered to according to consumer expectations, which explains the challenge of making significant changes to them (Celhay and Trinquecoste 2008). Within this relatively constrained framework, French winemakers highlight their expertise, drawing upon established techniques for vineyard management, winemaking and respect for the terroirs. Furthermore, collective strategies have emerged based on a close identification between territory, company and product, centered on the creation of a "territorial brand", generally consolidated by an appellation d'origine contrôlée (AOC). These initiatives enable them to compete in a highly globalized and fragmented market where individual commercial brands cannot sustain themselves as entirely independent entities due to limited budgets[7].

Box 6.2. *Summary of the "Rencontres nationales des Vignerons indépendants" (National Meeting of Independent Winemakers) Round Table, April 14, 2016*

6.3. Mental imagery: a close-up of its characteristics and how to use it wisely

While the value of employing mental images can be seen as an established fact in marketing, it still needs to be confirmed in the specific field of wine communication.

6.3.1. *General characteristics and applications of mental images in marketing*

> The imaginary relish is so sweet, that it enchants my sense (Shakespeare (1602), *Troilus and Cressida*, Act 3, Scene 2).

Mental imagery, which can be traced back to the earliest mentions by Plato and Aristotle, is undoubtedly one of the most well-known inner experiences. It is something each and every one of us encounters to varying degrees of frequency and

7 Excerpt from the discussions that took place during the roundtable session at the *Rencontres nationales des Vignerons indépendants* (National Independent Winemakers Meeting) on April 14, 2016. Available at: https://www.vitisphere.com/actualite-82479-entre-marque-collective-et-marque-privee-que-choisir8201.html.

awareness during everyday cognitive tasks or when recalling memories of past moments. The famous episode of Proust's madeleine anecdote, described in his work "*In Search of Lost Time*", exemplifies mental imagery as a catalyst for recalling all of the information stored deep within our memory since birth. Proust began with a present sensation – the aroma of a madeleine dipped in tea – and used it to immerse himself in a series of mental images from his childhood.

Much like Proust, we occasionally find ourselves consciously revisiting a scene in our minds, experiencing its sensations, smelling its scents, tasting its flavors or hearing its sounds, all without the presence of the source of these internal representations; in other words, without the direct stimulus that initially stored this information. This unique yet common experience is fundamentally different from dreaming because we remain awake and aware while forming mental images (White and Hardy 1998). During this immersive experience, we can fully engage in it even as we perform unrelated tasks like taking an elevator, walking down the street or even driving. Mental imagery is thus described as a "[…] quasi-perceptual experience; it resembles perceptual experience, but occurs in the absence of the appropriate external stimuli"[8]. In marketing, mental imagery is described as the mental representation of a constructed perception or sensation, akin to an almost perceptual experience (Gavard-Perret and Helme-Guizon 2003). However, it takes place in the absence of the original perceptual stimulus. Additionally, mental images are closely linked to the concepts of stimulus exogeneity and endogeneity. In the former case, the stimulus originates from the external environment (sounds, events, etc.) and triggers mental images in an individual. In the latter case, it reflects the individual's internal state of the stimulus (imagination, fantasy, memory, etc.), which is also a powerful source of generating mental images (Roeckelein 2004).

More specifically, when we evoke the memory or imagination of an object, service or idea promoted through communication vectors, we essentially "bring to life in our minds" one or more attributes intimately associated with the stimulus. For instance, this could be the recollection of an event or a situation triggered by a "print" advertising message, packaging or even the ambiance of a retail environment. Operating both as a process of mental visualization and as its outcome (the mental images), mental imagery is nevertheless subject to individual differences such as imaging capacity and individuals' processing styles (Gavard-Perret and Helme-Guizon 2003).

Another characteristic of mental imagery is its ability to encompass one or more sensory modalities, whether similar to or different from those of the percept (McInnis and Price 1987). For instance, viewing a "print" advertisement for ice-cold

8 Stanford Encyclopedia of Philosophy 2021. Available at: https://plato.stanford.edu/entries/mental-imagery/.

beer (visual percept) or hearing the sound of beer being poured into a glass (auditory percept) in a radio[9] commercial can evoke a taste of beer in the mouths (taste mental image) of exposed individuals. While mental images can take on various forms, including auditory, olfactory, gustatory or *kinesthetic*[10], the visual modality remains predominant, as demonstrated by Gavard-Perret and Helme-Guizon (2003). Furthermore, mental imagery possesses multiple dimensions, as detailed in Box 6.3.

– *Vivacity*: refers to the intensity with which a subject experiences mental images (Ellen and Bone 1991).

– *Clarity*: refers to the degree of precision and detail of the mental images evoked (Ellen and Bone 1991).

– *Ease of appearance*: the effort required to activate information associated with the stimulus, previously stored in memory (Ellen and Bone 1991).

– *Quantity*: the number of images created during information processing (Ellen and Bone 1991). The quantity and vividness of imagery are not related, as one individual can evoke a single, very clear and detailed mental image, while another may develop a multitude of pale, blurry and fleeting images (Helme-Guizon 1997).

– *Valence*: refers to the pleasant or unpleasant nature of the mental image evoked by a stimulus (McInnis and Price 1987).

– *Elaboration*: refers to the activation of information beyond what is strictly provided by the stimulus (Babin and Burns 1998).

– *Self-referentiality*: a particular form of elaboration, it represents the ability of mental imagery to include ourselves in the mental scenario (Ellen and Bone 1991).

Box 6.3. *The key dimensions of mental imagery used in marketing*

While the concept of mental visualization has primarily been studied in psychology and cognitive science, numerous studies have established its relevance in the field of marketing. Indeed, various research efforts have shown that mental images exert an influence on memorization, attitudes and intentions. MacInnis and Price (1987) pointed out that the interest in mental imagery in marketing arises from its potential to explain and predict phenomena, especially within the realm of consumer behavior.

9 Radio is an authorized medium for wine-related advertisements, but specific time slots have been established by decree. On Wednesdays, advertisements are permitted from midnight to 7 a.m., while on other days, they are allowed from midnight to 5 p.m.

10 Imagination of sensations.

The effectiveness of mental visualization has been particularly demonstrated in advertising contexts, where mental images, influenced by advertising stimuli, have the ability to *convey* the impact on consumer responses (Gavard-Perret and Helme-Guizon 2003). For example, we can observe that the congruence between elements in an advertisement (Helme-Guizon 1997), or the interaction of textual and visual elements in a mixed stimulus that illustrates the brand name and the product in a single integrative image, generates mental images that are conducive to enhancing memorization (Lutz and Lutz 1978).

Just as a simple iconic stimulus can influence mental images and attitudes towards the brand, Babin and Burns (1998) emphasized that portraying a product in *actual use* in an advertisement can trigger mental visualization, which in turn affects consumers' attitudes towards the advert and the brand. Similarly, visuals with explicit verbal messages (vs. implicit messages) stimulate intense mental imagery, capable of fostering a positive attitude and a favorable intent to purchase (Rossiter and Percy 1980). Mental visualization has also been examined in the context of radio advertisements, demonstrating that the addition of sound effects consistent with the promoted product[11] (vs. an oral message) leads to more vivid mental imagery and a more positive attitude towards the aired advert (Miller and Marks 1992). Moreover, individuals exposed to an advertisement can be encouraged to engage in mental imagery. Bone and Ellen (1992) found that prompting consumers to imagine themselves using the product generates more extensive and impactful mental visualization, thereby fostering a more favorable attitude towards the advertisement.

Marketing research has shown that mental imagery acts *as a catalyst* that bridges the gap between iconic and verbal stimuli and individual responses. While mental visualization influences personal attitudes and intentions, there is a specific imaging technique known as "instructions to imagine" that appears to significantly enhance mental visualization and, as a result, the attitudinal and intentional responses it generates.

6.3.2. *Advertising and the Évin law: suggest rather than show, the Bordeaux wine example*

Mental Imaging Instructions (Imagery Instructions), also referred to as *Mental Imaging Scripts,* involve providing individuals with direct guidance to create a mental "visualization". Encouraging consumers to think about a specific object, prompting them to imagine themselves in a particular situation or leading them,

11 Refer to the beer advert example described above.

either directly or indirectly, to envision themselves consuming the promoted product can all be categorized as mental imagery instructions.

Previous research has shown that the presence of instructions to be imagined, included in a communication, results in a higher impact on the dimensions of quantity, vivacity and elaboration of mental imagery (see Figure 6.2), compared to the same advertisement not containing instructions (Babin and Burns 1998). It therefore seems that inducing or encouraging the consumer, exposed to a press or display type advertisement, to form mental images relating to the promoted product makes the mental viewing experience more intense, as the instructions affect certain particular dimensions of the imagery.

In terms of individual responses to mental imagery induction scripts, other studies have also emphasized the positive impact of instructions. This impact extends to attitudes towards advertising and the presented brand (Babin and Burns 1998; Bréhonnet and Lagier 2019). Therefore, it is well established that encouraging consumers through advertising to mentally visualize the promoted product and a pleasurable use of it can significantly influence attitudes and behavioral intentions.

Certain dimensions of mental imagery can account for some or all of the effect of an advertising stimulus on the attitudinal and behavioral responses of exposed consumers. In reality, the positive effect does not stem from mental imagery as a whole, but rather from specific dimensions of it (see Box 6.3). These dimensions play a *pivotal role* (or mediating role[12]) in this stimulus-response relationship.

This *specific mediating role* assigned to certain dimensions of mental images involves conveying and/or amplifying the measurable (visible) effect of a stimulus on attitudes and behavioral intentions.

Babin (1992) demonstrated, for instance, in an advertising context, that the quality and elaboration of imagery contribute to the formation of attitude towards the brand and the advert. The vividness of imagery has also been found to mediate the effect of an advertising message containing a concrete verbal message on attitudinal responses (Burns et al. 1993). In the context of France, Helme-Guizon (1997) explored multisensory mental imagery, revealing the mediating role of certain dimensions on individual responses.

In an advertising context where situations are designed to encourage the activation of mental imagery, Babin and Burns (1998) also found that the vividness

12 Mediator variables are intermediate variables (located between the stimulus and the response) that explain the relationship between the stimulus and an individual response.

and elaboration of the evoked mental imagery act as mediators in the impact of the instructions to imagine on attitudes towards the brand and the ad. However, these authors do not provide evidence of any role for the quantity of imagery. Although the same research indicates that this dimension appears to have a mixed role in terms of mediating effects, it is worth noting that several other studies emphasize its significance. Ellen and Bone (1991) reveal that a stimulus containing instructions to imagine influences both the quantity and ease of imaging. On the other hand, Miller et al. (2000) demonstrate an effect of imagery quantity on attitudes towards advertising and the brand.

In summary, the research discussed earlier confirms that there are several advertising strategies capable of enhancing individuals' mental imagery, namely the *use of images or sounds*, as one approach, and the *inclusion of imagination instructions* in advertisements, as another. Some studies highlight that this mental visualization ultimately has the potential to influence attitudinal judgments, specifically by *mediating* them. Furthermore, other studies demonstrate that advertisements containing imagination instructions also stimulate mental imagery, consequently leading to a significant enhancement of attitudes.

It is worth mentioning that the sensitivity of mental image dimensions and individual responses to imagination instructions has been examined in numerous marketing studies. Some of these are outlined in Table 6.1.

In the following sections, we have opted to provide a more detailed presentation of the results from a recent study on this topic (Bréhonnet and Lagier 2019). This research specifically delved into the viticultural context of Bordeaux wines and examined the impact of imagination instructions integrated into a print advertisement on individual responses. The experiment strictly adhered to the terms outlined in the 2015 amendment to the Évin law. As we will explore in more detail later, the findings are likely to demonstrate to practitioners *that skillfully encouraging* individuals to envision the terroir's benefits with *carefully chosen* words can significantly influence attitudinal responses towards the underlying product.

The experiment employed a straightforward advertising stimulus, created from a photograph depicting the landscape and culinary elements associated with a production terroir, similar to the Bordeaux terroir, for example. The visual features red grapes and vine leaves, with an unidentifiable wine bottle on the right side and a glass containing the product in the foreground (see Figure 6.1).

Stimulus characteristics	Imaging dimensions	Studied responses	Authors
Imaging instructions involve providing directions to the subject to create mental images associated with the concept to be retained.	Propedeutics of mental imagery	– Mental imagery – Memory formation	Lutz and Lutz (1978)
Advertisements with or without imagination instructions	Vivacity, quantity, elaboration	– Mental imagery – Attitude towards advert/brand – Imaging mediation	Babin and Burns (1997)
Radio sound effects, vivid verbal messages, imagination instructions	Vivacity, quantity	– Mental imagery – Feelings about the advert – Attitude towards the advert	Miller and Marks (1992)
Advertisements with or without imagination instructions	Quantity, vivacity, elaboration	– Mental imagery – Attitude towards the advert/brand	Babin and Burns (1998)
Wine advertising with or without imagination instructions	Quantity, vivacity, elaboration	– Mental imagery – Attitude towards the ad/brand	Bréhonnet and Lagier (2019)

Table 6.1. *Key marketing studies examining the impact of imagination instructions on the dimensions of mental imagery and individual responses*

Figure 6.1. *Advertising stimulus without imaging script (photo credit © Bréhonnet R.). For a color version of this figure, see www.iste.co.uk/cheriet/wine2.zip*

A second visual, identical to the first, was used, but this time it included a script encouraging individuals to imagine the terroir and its attributes: “Imagine the sweetness of the terroir... Imagine on your palate its fruity flavor... Imagine its gentle warmth within you...” (see Figure 6.2).

Figure 6.2. *Advertising stimulus with imaging script to imagine the terroir and its attributes (photo credit © Bréhonnet R.). For a color version of this figure, see www.iste.co.uk/cheriet/wine2.zip*

A first group of individuals (control group) was exposed to the visual without instructions to imagine the terroir (see Figure 6.1), while a second group (test group) was exposed to the visual containing explicit instructions to imagine the terroir and its attributes (see Figure 6.2). The impact of these instructions on the typical dimensions of elaborated mental imagery, as well as on attitudes towards the advert and the brand (assumed in this study to be “Bordeaux wines”), was examined and assessed using psychometric scales that had been previously validated in other studies (Bréhonnet and Lagier 2019).

In the context of wine communication, the study reveals that the presence of instructions to imagine in the “print” visual more strongly encourages an individual to mentally create a consumption experience, even if it is not explicitly prompted. This aspect of mental imagery remains the most responsive to imagery scripts and the most prominent when this specific technique is used. Consequently, exposing the participant to instructions to imagine the terroir with “carefully chosen” words can lead them, often unconsciously, to form mental images that extend well beyond the elements of the terroir shown in the visual. Although the imagery instructions used in the visual do not strictly encourage a literal mental tasting of the wine, the power of this suggestion lies in the skillful use of words integrated into the imagery script, with the phrasing aligning with the iconic elements representing the terroir. In this very particular case, if the visual is well-crafted, a mental transfer of the terroir's elements to the promoted brand (wine) occurs. The exposed individual can then vividly experience multisensory mental images (the taste of grapes, the warmth of the vine, the flavor of the beverage), encouraging the exploration of the product.

The study also demonstrates that participants exposed to the visual containing imaging instructions develop a significantly more favorable attitude towards the advertising visual and the wine brand, in contrast to a visual lacking imaging instruction. Considering that desire lies mentally between attitude and the actual act of consumption (Perugini and Bagozzi 2001), this finding implies that the generated attitude is more likely to translate into a concrete action of tasting and purchasing.

These results carry significant implications, as they distinctly demonstrate the positive influence of the instructions to imagine on attitudes towards wine brands and advertising. They also underscore the role played by the development of mental imagery in this context. Additionally, it is noteworthy that the encouragement to imagine is related to the characteristics of the region and its products, which appears to align with the 2015 amendment to the Évin law, making such advertising legally acceptable.

6.4. Conclusion

As part of its most recent amendment, the Évin law provides a less restrictive arrangement for sectoral communication on wine.

However, it obviously continues to require advertisers and agencies to adapt strategies, messages and media as a result of its new provisions.

For advertisers and agencies, one effective adaptation strategy involves using mental imagery: a technique that has been consistently shown to enhance attitudes towards products that employ it, or at the very least, towards the contexts they evoke. This effect is particularly expressed in terms of valence, signifying the more or less pleasant nature of the suggested mental image.

The potential use of mental imagery is not merely a theoretical or "virtual" opportunity; it is a practical reality. Several advertisers have already employed it, and its effectiveness has been assessed in numerous prior studies.

In this regard, these previous studies lead advertisers in the industry to two interesting conclusions: firstly, the value of taking inspiration from a form of communication considered legal to effectively serve their ongoing goals, and secondly, the importance of analyzing the performance of such communications in terms of attitude and actual behaviors, provided that this information can be known and collected.

6.5. References

Babin, L.A. (1992). Effects of imagery-eliciting strategies on imagery processing memory, beliefs, attitudes, and intentions from print advertising. PhD Thesis, Louisiana State University, Baton Rouge.

Babin, L.A. and Burns, A.C. (1997). Effects of print ad pictures and copy containing instructions to imagine on mental imagery that mediates attitudes. *J. Advert.*, 26(3), 33–44.

Babin, L.A. and Burns, A.C. (1998). A modified scale for the measurement of communication-evoked mental imagery. *Psychol. Market.*, 15(3), 261–278.

Bagozzi, R.P. and Perugini, M. (2001). The role of desires and anticipated emotions in goal-directed behaviors: Broadening and deepening the theory of planned behavior. *Br. J. Soc. Psychol.*, 40, 79–98.

Beverland, M. and Luxton, S. (2005). Managing integrated marketing communication (IMC) through strategic decoupling: How luxury wine firms retain brand leadership while appearing to be wedded to the past. *J. Advert.*, 34(4), 103–116.

Bone, P.F. and Ellen, P.S. (1992). The generation and consequences of communication evoked imagery. *J. Consum. Res.*, 19(1), 93–104.

Bréhonnet, R. and Lagier, J. (2019). Quand "l'instruction à imaginer" fait évoluer la communication sur les vins. *Décisions marketing*, 93(1), 53–70.

Burns, A.C., Biswas, A., Babin, L.A. (1993). The operation of visual imagery as a mediator of advertising effects. *J. Advert.*, 22(2), 71–85.

Celhay, F. and Trinquecoste, J.F. (2008). Pourquoi est-il difficile de "sortir" des codes graphiques du vin français ? *Market management*, 8(1), 6–30.

Château, L. (2009). La Dive bouteille en France : une analyse sémiologique et linguistique de son discours et de ses images. *COnTEXTES*, 6 [Online]. Available at: https://journals.openedition.org/contextes/4485.

Croijmans, I., Speed, L.J., Arshamian, A., Majid, A. (2019). Measuring multisensory imagery of wine: The vividness of wine imagery questionnaire. *Multisens. Res.*, 32(3), 179–195.

Dufour, S. and Boutaud, J.J. (2012). Le sacré et son terroir. De la communication des maisons de vin et de champagne. In *Entreprise et sacré. Regards transdisciplinaires*, Boutaud, J.J. (ed.). Hermès Lavoisier, Paris.

Duquesnois, F. and Cheriet, F. (2021). Loi Évin, marketing du vin et protectionnisme. *Open Wine Law*, 1(1), 1–17 [Online]. Available at: https://openwinelaw.fr/article/view/4747.

Ellen, P.S. and Bone, P.F. (1991). Measuring communication-evoked imagery processing. *Adv. Consum. Res.*, 18, 806–812.

Gallopel-Morvan, K., Spilka, S., Mutatayi, C., Rigaud, A., Lecas, F., Beck, F. (2017). France's Évin law on the control of alcohol advertising: Content, effectiveness, and limitations. *Addiction*, 112, 86–93.

Gavard-Perret, M.L. and Helme-Guizon, A. (2003). L'imagerie mentale : un concept à (re) découvrir pour ses apports en marketing. *Recherche et applications en marketing*, 18(4), 59–79.

Helme-Guizon, A. (1997). Image, imagerie mentale et effets de la communication persuasive : application à une œuvre d'art incluse dans une annonce publicitaire. PhD Thesis, Université Paris-IX Dauphine, Paris.

Lutz, K.A. and Lutz, R.J. (1978). Imagery-eliciting strategies: Review and implications of research. *Adv. Consum. Res.*, 5, 611–620.

MacInnis, D.J. and Price, L.L. (1987). The role of imagery in information processing: Review and extensions. *J. Consum. Res.*, 13(4), 473–491.

Miller, D.W. and Marks, L.J. (1992). Mental imagery and sound effects in radio commercials. *J. Advert.*, 21(4), 83–93.

Miller, D.W., Hadjimarcou, J., Miciak A. (2000). A scale for measuring advertisement evoked mental imagery. *J. Mark. Commun.*, 6, 1–20.

Mora, P. and Livat, F. (2013). Does storytelling add value to fine Bordeaux wines? *Wine. Econ. Policy.*, 2(1), 3–10.

Passebois-Ducros, J., Trinquecoste, J.F., Pichon, F. (2015). Stratégies d'artification dans le domaine du luxe : le cas des vins de prestige. *Décisions marketing*, 109–124.

Roeckelein, J.E. (2004). *Imagery in Psychology: A Reference Guide*. Greenwood Publishing Group, Westport.

Rossiter, J.R. and Percy, L. (1980). Attitude change through visual imagery in advertising. *J. Advert.*, 9(2), 10–16.

7

Determining the Value of Organic Options through Willingness to Pay

7.1. Introduction

Due to its history and tradition, wine holds a unique position on the French consumer's table. Nevertheless, consumption habits continue to evolve. While the overall quantity of consumed wine has decreased (-0.8%), its monetary value has increased (1.2%). Between 1980 and 2015, the percentage of regular consumers decreased from 47% to only 16.4%, while occasional consumers increased from 30% to 50%, and non-consumers saw their numbers rise from 23.6% to 32.8% (FranceAgriMer 2019, 2020).

Wine has become a product that is increasingly associated with special occasions and is less commonly consumed as part of daily meals. The pursuit of a healthy diet plays a significant role in the decline of wine consumption, as it is perceived as detrimental to health, due to its alcoholic nature (Hill and Laplanche 2010; Hashibe et al. 2016)

This is why the research presented in the following pages specifically examines the role of the organic attribute (bio) in enhancing the perception of wine among consumers.

7.2. Consumers' willingness to pay for organic wine

Despite the decrease in purchasing power, 36% of French consumers consider it reasonable to pay more for organic food. Reasons cited include health benefits

Chapter written by Alvaro CUYA GAVILANO and Jean-François TRINQUECOSTE.

(29%), environmental benefits (12%), improved product taste (9%) and ethical concerns (8%). All of these factors contribute to consumers showing a willingness to pay a premium for organic products. However, this willingness is not without its limits. In 2020, the primary reason for consumers not purchasing organic products was, by a significant margin, the perceived high cost (73%) (AgenceBio 2021).

Indeed, organic products are typically 20–30% more expensive than non-organic products. Therefore, it is not surprising that consumers place importance on this price difference, especially in competitive markets such as the agricultural sector. This is particularly true in the case of the wine market in France, where there is significant competition from high-quality non-organic wines (Jonis et al. 2008), including those with a controlled designation of origin.

From producers' perspective, the higher price is justified by the difference in production costs and the additional quality they believe the product possesses. However, it is essential to confirm whether consumers themselves perceive and value this added quality in organic wines. In essence, the question to address is: Does the organic attribute increase the perceived value of wine in the market and consequently contribute to an increased willingness to pay (WTP[1]) among at least some of the buyers for such products? Previous research suggests this possibility. Beck et al. (2012), for instance, noted that "consumers prefer and purchase organic foods because they believe they are of higher quality", and therefore "they are also willing to pay a premium price". Additionally, some authors argue that consuming organic products reflects consumer values (Thøgersen 2011). The study presented below aims to empirically investigate the validity of these claims.

7.2.1. *Definition of WTP*

In general, WTP represents the "maximum amount" a consumer is willing to pay. When a consumer agrees to pay the price requested, they only partially express their perception of the value of the product because their WTP may be equal to (or higher than) the market price. This is why WTP can be mistaken for the *reservation price*, which is the price at which the consumer is indifferent between "consuming" and "not consuming" (Breidert 2006; Le Gall-Ely 2009).

For example, if a consumer considers that a wine deserves to cost 50 euros, and buys it from their wine shop at 30 euros, they do not really reveal their WTP; the situation allows them to implicitly keep their reservation price, which is defined as

---

1 The reader should note that the acronym WTP will be used to express the concept of "willingness to pay" throughout the rest of this chapter.

the difference between what they would agree to pay to acquire the product and what is actually asked of them to do so.

Consider this other example: a consumer regularly buys a wine for 20 euros from their wine merchant. On a trip to New York, they would like to introduce this wine to their friends. They found it, but for 60 euros. "Too expensive," they told themselves, "if it had been 40 euros, I would have bought it". This situation shows that the *maximum price* they would have been willing to pay is 20 euros higher than what they regularly pay. The current price at the wine merchant is referred to as the *reference price*, and the additional 20 euros they were willing to pay represents the *price difference* they would have considered acceptable. The maximum price is defined as the reference price, plus this difference.

From these considerations, we can derive two key insights for businesses. The first insight pertains to the importance of understanding the factors that influence consumers' WTP. The second insight is about the importance of correctly setting prices relative to the consumer's WTP, which involves assessing the value of this WTP, both before purchase and after use or consumption. Ultimately, these two values determine the short- and medium-term demand volume. Understanding the determinants of WTP theoretically gives meaning to its measurement, and from a managerial perspective, it enhances segmentation, targeting strategies and the overall marketing mix, including pricing strategies.

7.3. Personal determinants of WTP

It is important to acknowledge that consumers are not solely "maximizers" of economic utility, but also citizens. This implies that the consumption of organic food can be viewed from two perspectives: individual consumerist motivations and citizen motivations. Individuals sometimes act primarily as consumers driven by self-centered motives and personal interests. Conversely, when faced with decisions that have significant environmental impacts, they may act as "responsible" citizens and prioritize a societal perspective (Sagoff 2008).

7.3.1. *Personal values and WTP*

Individual behavior is directed towards more or less abstract goals, such as their personal values, which are expressed in daily life. For instance, when deciding between organic and non-organic wines at the supermarket, the individual has to choose between individual and collective goals. According to Schwartz and Bilsky (1992), values are "concepts or beliefs about desirable end states of existence that transcend specific situations, guide the choice or evaluation of behaviors and events, and are organized in an order of relative importance" (Schwartz et al. 2012).

Pro-environmental behaviors, such as consuming organic food, can be seen as mixed behaviors, characterized by both individual and collective motivations. Altruistic motivations involve a sense of moral obligation towards society, the environment, wildlife, etc. Typically, environmental, social and ethical values are linked to purchasing "green" products. However, some authors argue that individualistic values, such as safety, health and hedonism have the most significant influence on its consumption (Joshi and Rahman 2015).

Consumers who are primarily motivated by the pursuit of pleasure and growth, in other words, hedonistic motivations, are likely to be especially drawn to products perceived as having better taste quality. In this regard, several studies have demonstrated a positive relationship between the organic attribute and improved perceived food quality (Fotopoulos et al. 2011).

Likewise, wine consumption is linked to the pursuit of pleasure and conviviality (Remaud and Sirieix 2012). Consequently, it would appear reasonable to anticipate a positive relationship between the WTP for a wine perceived as of higher quality, and values such as hedonism, stimulation and even autonomy.

However, and this is a crucial point, organic wines have not been linked to wines of superior quality. Sometimes, an inverse relationship is even considered more common. For instance, the study conducted by Remaud and Sirieix (2012) reveals that organic wine "is not associated at all with a good quality/price ratio". This situation is likely to negatively influence the WTP assessment for this type of wine. Conversely, when consumers perceive that they must choose between products that are identical in terms of sensory and non-sensory attributes, we can expect a positive correlation between values associated with openness to change and the WTP for wine with the organic attribute.

Studies and consumption statistics also indicate that the choice of organic foods is motivated by the pursuit of health benefits. However, as known, while wine can be considered a relatively healthier and more natural beverage than other highly alcoholic drinks, it is not typically consumed with the aim of improving the health of the person drinking it.

7.3.2. *WTP as behavioral intention*

Wine can be seen as a product with a high level of involvement, and selecting it may require more "self-control" compared to other everyday products with lower involvement. In most instances, purchasing wine is not typically characterized as impulsive behavior, but rather as a thoughtful decision. This holds particularly true when choosing a wine that carries the organic label.

While choosing and assessing a wine can be viewed as rational behaviors, it is important to understand them beyond the explanations provided by the "neoclassic" approach, which focuses on maximizing individual economic benefits. Firstly, wine is a product associated with tradition, the land, as well as conviviality, socialization and aesthetics – aspects that extend beyond purely economic considerations. Secondly, individuals in question are not isolated entities or consumers but integral parts of a social community.

Furthermore, voluntary purchasing actions are characterized by both conscious and unconscious goals and processes. Voluntary behavior results from the processing of available information and context. Behavioral intention is the fundamental element in explaining this type of behavior. Two major theories are particularly aimed at explaining this behavioral intention: firstly, the theory of action phases (Achtziger and Gollwitzer 2008); and secondly, the theory of planned behavior (Ajzen and Dasgupta 2015).

Based on the "action phases" model, a consumer establishes their goals by assessing various alternatives, based on their attractiveness (values) and feasibility. Consumers initially define their objectives, and then determine when to "take action". When a buyer chooses to pay for an organic wine, they are striving to achieve both individual and collective goals, which can sometimes be partially contradictory.

According to the theory of planned behavior, behavioral intent is the immediate precursor to any voluntary action. Intent signifies a disposition to act in a certain manner, in specific circumstances. Behavioral intent serves as a mediating factor that bridges the gap between attitudes formed by beliefs and the resulting behavior. Within this process, subjective norms come into play, based on normative beliefs, as does perceived behavioral control, which reflects beliefs about individual constraints or barriers (economic, material, etc.), along with the context (product availability, quantity, price, etc.) (Ajzen and Dasgupta 2015).

As mentioned earlier, when a consumer decides to pay the price offered by the company, it does not necessarily reveal their WTP. At most, it can be inferred that their WTP is equal to (or greater than) the price paid. Directly asking the consumer what their WTP is only provides access to the declared WTP, but not necessarily the true WTP. For this reason, WTP, as the maximum price, is more associated with the intent to act than with actual behavior. There are several applications of the concept of behavioral intent that exist to explain and provide substance to monetary measures of WTP.

7.4. Empirical analysis of WTP

From a methodological standpoint in marketing, intent is typically measured using psychometric scales, which can potentially disconnect the measurement from its actual monetary reality. To overcome this limitation, various methods have been employed in previous research, including contingent valuation, hedonic pricing and other econometric approaches. In marketing, preferred methods often involve conjoint analysis and other survey-based techniques (Le Gall-Ely 2009). These methods are characterized by an inherent information asymmetry, as individuals are reluctant to disclose their true WTP to avoid potentially paying more.

The research presented in this text used three complementary approaches: the first aimed to study WTP determinants, based on the measurement of consumers' psychometric constructs; the second aimed to propose an indirect WTP measurement, through a discrete choice experiment; and the third sought to establish a direct and incentivized WTP measurement, using a pseudo-auction method reminiscent of the Becker–DeGroot–Marschak (BDM) approach. The first approach allows for the examination of personal WTP determinants. However, it does not enable its monetary measurement or the assessment of sensory attributes; hence, the need to complement it with an experimental approach, including a tasting session in the tasting room.

The experiment was divided into three parts: the first part involved asking participants to complete a questionnaire to measure psychometric constructs; the second part was a discrete choice experiment, where participants had to make choices among different wine profiles; and the third part involved organizing an auction with a real commitment to purchase.

7.4.1. *Observation of consumer preferences and WTP*

There is a significant challenge to measuring WTP: how to measure the potential gap between what an individual truly thinks and what they declare they are willing to pay. The person being interviewed may act as a free-rider by underestimating the actual amount they are willing to pay. Conversely, individuals confronted with the question "How much are you willing to pay for products that support the planet?" may declare an amount higher than their actual WTP to conform to social expectations (social desirability bias).

Some authors measure WTP by directly asking consumers open-ended questions like "How much are you willing to pay?". Others prefer not to ask the question directly but opt to indirectly infer WTP from consumer choices, for example, choices expressed through an experimental design, based on conjoint analysis. In practice, various empirical studies demonstrate that both approaches (direct and indirect) can yield inaccurate results due to psychological biases and other technical factors. The fundamental benefit of a context-based measurement approach is to involve the respondent in a process that exposes them to real consequences, meaning that participants must genuinely purchase the study product.

7.4.2. *Study field and sample*

The study field was determined by the interest expressed by the Bordeaux region, which funded this study, in the wine industry. It should be explicitly mentioned that this funding, while enabling the research to be conducted, had no influence whatsoever on the choice of methodology adopted by the researchers or, more importantly, on the results obtained. With this clarification in mind, it is worth noting that, in terms of the research method, it was decided to conduct an exploratory study using the "snowball" technique with a sample of 285 wine consumers. The prerequisite for inclusion in the sample of respondents was being a wine consumer, specifically having consumed wine in the month preceding the study.

For the main study, a sample of 148 respondents was assembled. To achieve this, the following selection criteria were applied: (1) being a wine consumer (at least once a month); (2) being under the age of 60 years; (3) preferably employed; (4) not working in the wine industry; (5) not having participated in any similar studies (e.g. marketing, tasting, etc.) in the last three months. A specialized company in consumer panel recruitment was enlisted for this purpose. The core of the research focused on an active population, aged between 22 and 60 years. The sample was balanced in terms of gender.

7.5. Study 1: WTP determinants

7.5.1. *Personal values and behavioral attitude*

This study aimed to assess the influence of values on attitudes towards purchasing and the willingness to pay a premium for organic wine. The nature and intensity of these attitudes were explicitly linked to three categories of personal values, namely: "values associated with conservation", "values associated with openness and change" and "self-enhancement". These categories were derived from the 10 Schwartz values. We conducted the necessary statistical tests and verifications, and employed a robust regression based on the weighted least squares method. The results are presented in Table 7.1.

From this study, it can be concluded that factors associated with "conservation-oriented" values, such as safety, tradition and conformity, have a negative influence on attitudes towards purchasing (β = – 0.10) and the willingness to pay more for organic wine (β = – 0.19). This result is noteworthy because, in most studies on the consumption of organic foods, the primary motivations typically revolve around health and nutrition, specifically aimed at preventing "losses" and threats.

The specific case of the attitude towards paying more for organic wine can be explained by the predominance of the perceived financial risk, that of paying a higher price for organic wine that may not necessarily be better than non-organic products. In other words, consumers appear to prioritize the disadvantages of paying more today over the potential but long-term benefits to their health.

Dependent variables	Intercept	Conservation	Openness to Change	Self-enhancement
Attitude to buy	3.718**	*– 0.1079**	*0.2645***	0.0482
Attitude to pay	2.976**	*– 0.1921***	0.1725	*0.1883***

* Indicates significance at the 10% level.
** Indicates significance at the 5% level.

Table 7.1. *Regression results; personal values versus behavioral attitude*

It is also noteworthy that openness to change (i.e. universalism, hedonism and autonomy) has a positive influence on consumers' attitude towards purchasing organic wine (β = 0.26). This aligns with previous research that links universalism positively to the consumption of organic foods. However, this openness to change does not have the same impact on the attitude towards paying more for organic wine. Consumers who are more open to change exhibit a positive attitude towards buying organic wine, but this attitude does not necessarily predispose them to accept paying a higher price.

Value(s) associated with self-enhancement (power and achievement) positively influence consumers' attitude towards paying more for an organic wine (β = 0.18). This result may be explained by the existence of a positive relationship between the personal image associated with being able to afford such a wine and, correspondingly, the fact that this type of purchase becomes a "marker" of personal success. However, they do not significantly increase the actual purchases of this type of wine.

7.5.2. *Structural model of WTP*

This model aims to explain two aspects: firstly, the purchase of organic wine, measured by the self-reported frequency of purchase, and secondly, the WTP assessed through the intention to pay an additional price for wine with the organic attribute. The model includes six explanatory variables: behavioral attitude towards purchasing organic wine, behavioral attitude towards paying a premium for organic wine, social pressure or peer influence measured by normative beliefs and subjective beliefs regarding the purchase of organic wine, and perceived behavioral control, which encompasses external barriers to paying more for organic wine and internal barriers related to perceived self-efficacy, measured by perceived knowledge level.

The exploratory analysis clearly indicates the presence of the six latent variables: attitude towards the idea of paying more, attitude towards the idea of purchasing, subjective norm, perceived behavioral control (related to payment), subjective knowledge about organic foods and subjective knowledge about wine. The confirmatory analysis is also satisfactory, enabling the analysis to proceed.

The overall fit quality of the model is satisfactory. As expected, the WTP for an organic wine significantly (and positively) influences purchase behavior (intention = 0.252): the more positive the WTP, the higher the probability of buying an organic wine increases. This underscores the importance for companies to justify the value of an organic wine.

The willingness to pay more for an organic wine is influenced by the favorable attitude towards such a purchase (AttPay = 0.530) and the willingness to accept paying more to access the product (AttBuy = 0.423). These findings emphasize the importance of not solely relying on the presence of a favorable attitude towards the hypothetical purchase of the product but also ensuring that the consumer has a positive attitude towards the idea of spending more to acquire it.

		Intention (pay)	**CCPPay**	**AttBuy**	**CSVin**	**CroNor**	**CSBio**	**AttPay**
Intention (pay)	**Coef.**	–	0.179	0.423	−0.074	0.111	0.185	0.530
	P-value	–	0.080	0.006	0.398	0.075	0.007	***[2]
Behavior	**Coef.**	0.252	–	–	–	–	–	–
	P-value	***	–	–	–	–	–	–

Table 7.2. *Structural model estimation*

2 *** Indicates significance at the 1% level.

Results also indicate an, albeit weak, influence of social pressure (CrNor = 0.111), as measured by the normative beliefs prevailing within the "relevant" social circle (family, friends, colleagues). This underscores the importance for companies to anchor their communication in the social acceptance of consuming organic wine, along with the "necessity" to pay more.

Results confirm the influence of the perception of control exerted by external barriers, such as product availability (CCPPay = 0.179), and also highlight the significance of the perceived knowledge level related to the product of interest (CSBio = 0.185). This finding suggests the potential implementation of a consumer education strategy. The present research demonstrates that subjective knowledge about organic food has a significant positive impact on the intention to pay more. However, the acquired and available knowledge about wine does not significantly affect the WTP assessment for an organic wine.

7.6. Study 2: indirect WTP measurement

In everyday life, consumers do not assess a product in isolation. When faced with what is on offer in a supermarket aisle, they comparatively evaluate multiple products, each with its own set of attributes. The concept of a "discrete choice" experiment is based on McFadden's theory (1974). This theoretical approach aims to explain choices made through multiple product comparisons, rather than solely evaluating the individual "qualities" of their attributes.

Consumers' choice is driven by a deterministic component, coupled with a *random component*. The deterministic component is a function that depends not only on the attributes of the product (e.g. color, appellation, label, production system, price) but also on the factors associated with the consumer's characteristics (e.g. psychological, social, economic).

7.6.1. *Protocol*

From an experimental perspective, we can distinguish between two types of attributes: *generic attributes* and *specific attributes*. Generic attributes are treated the same way for all products. A specific attribute, as the name suggests, is a characteristic not shared by the other products within the set of choices presented.

The analysis we conducted, as reported in the following lines, focuses on five attributes considered relevant for the French population. These attributes were determined based on previous research on wine consumption behavior in Europe and France. Furthermore, this database has been refined with input from consumers and experts gathered through interviews.

In this discrete choice experiment (see Table 7.3), participants are presented with a scenario where they must choose one wine from three distinct profiles labeled as A, B and C. This selection process is repeated 20 times, each time with different scenarios featuring various wine profiles. For each scenario, respondents are tasked with evaluating the expected quality of each wine, based on the provided information. The expected quality is assessed using a differential semantic scale, ranging from poor quality to good quality.

It is important to note that in certain situations, consumers are required to make a choice from a set where their favorite option is not included.

7.6.2. *Econometric estimation model*

Consumer choice has been modeled as a latent utility function, dependent on both the attributes of the wine and the consumer's characteristics. The presented scenarios require consumers to choose a bottle of wine based on non-sensory attributes and personal traits in two distinct situations: one involving domestic consumption and another, presumably less common, scenario of a gathering among friends. This duality in forecasted consumption situations aims to evaluate the impact of situational involvement.

We opted for a Logit model, where the dependent variable measures the probability of selecting a wine with the organic attribute (at home and with friends). The explanatory variables represent the non-sensory attributes of the wine, including color, AOC (Bordeaux, Bordeaux supérieur and Pessac-Léognan), vintage, production system and price.

Scenario 1					
A		B		C	
Appellation contrôlée Bordeaux supérieur		Appellation contrôlée Bordeaux		Appellation contrôlée Bordeaux supérieur	
Color:	White	Color:	White	Color:	Red
Year:	2007	Year:	2012	Year:	2007
Origin:	Conventional	Origin:	Conventional	Origin:	Conventional
Price:	€10.20	Price:	€4.20	Price:	€21.70
Based on this information, I think this wine is of		Based on this information, I think this wine is of		Based on this information, I think this wine is of	
Poor quality – Good quality		Poor quality – Good quality		Poor quality – Good quality	
0 1 2 3 4 5 6 7		0 1 2 3 4 5 6 7		0 1 2 3 4 5 6 7	

Table 7.3. *Example of a discrete choice experiment scenario plan*

It is quite common to emphasize that, in the vast majority of cases, consumer choice depends on the price of the product offered. Although wine has a multitude of characteristics, consumers often use price as a proxy variable for perceived quality, which significantly influences their choice. In other words, the higher the price of the wine, the more positively the consumer perceives the likely quality of the product. However, consumers do not always purchase the most expensive products, even when seeking reassurance about the quality of what they wish to acquire.

To isolate the dual role of price, we introduced perceived quality as an explanatory variable for the choice. Two models were simultaneously estimated: first, in a situation with a low level of involvement, the choice of a wine intended for ordinary and domestic consumption; second, in a situation with a high level of involvement, the choice of a wine for a more exceptional consumption, as planned during a gathering of friends who are connoisseurs and gastronomes.

Both models provide a good overall explanation of consumer choice. The models also have an acceptable predictive accuracy. It is worth noting that our research primarily aims to contribute to understanding consumer choice. The coefficients of a Logit regression are not directly interpretable. Therefore, to analyze the influence of each variable, it is important to measure marginal effects (refer to Table 7.4).

Drinking occasions		**Color**	**Bdx_Sup.**	**Pessac**	**Year**	**Sys._prod.**	**Prix_EU**	**Quality**
Home	**dp/dx**	– 0.011	0.111	0.113	0.000	– 0.014	– 0.019	0.066
	p-val	0.316	0.000	0.000	0.993	0.246	0.000	0.000
Friends	**dp/dx**	0.001	0.199	0.204	– 0.010	*0.058*	– 0.004	0.089
	p-val	0.920	0.000	0.000	0.000	*0.000*	0.000	0.000

Table 7.4. *Marginal effects: consumption at home and gathering with friends*

The choice related to consumption among friends leads consumers to evaluate a greater amount of information than that related to domestic consumption. Consequently, the number of significant estimated coefficients increases (such as vintage and production system), compared to the model for home consumption. Moreover, the other variables that were significant for "usual and domestic" consumption, including AOC, price in euros and expected quality, remain significant for consumption with friends. Therefore, a social consumption situation requires evaluating more attributes than a situation with a low level of involvement. This finding confirms that the consumption context influences both the number and the nature of the selection criteria employed by consumers.

In line with previous research findings, the Appellation d'Origine Contrôlée (AOC) positively influences consumer choices. Following the general law of price elasticity of demand, the estimated coefficient has a negative impact on the probability of selecting a wine. However, this marginal effect is less pronounced in social consumption situations, compared to the choice of wine for home consumption. In other words, consumers are less price-sensitive when choosing wine for social gatherings with friends. Additionally, it is noteworthy that a 1% improvement in expected quality increases the probability of wine selection by 6.6–8.8%.

When it comes to purchases meant for regular domestic consumption, the influence of the vintage is negligible. However, in the context of buying wine for social gatherings with friends, a more recent vintage is associated with a lower likelihood of being chosen.

Another significant difference in the results, depending on the consumption context, relates to the impact of the production system (organic vs. conventional). The organic attribute appears to have no influence on the choice of wine for regular domestic consumption. Conversely, in the case of social consumption, when a consumer is comparing two similar wines, the presence of the organic attribute increases the probability of choosing a wine by nearly 6%.

Our model enables the measurement of the financial "sacrifice" that consumers are willing to make in order to obtain a specific attribute. The WTP can be viewed as the ratio of the coefficient of the non-monetary attribute of the product (production system, AOC, etc.) to the coefficient of the price, all else being equal (ceteris paribus).

Calculating the WTP for the organic attribute in a "usual and domestic" consumption situation is impossible. The reasons for this may be multifaceted. It is possible that in this scenario, the WTP is close to zero, indicating that consumers are not willing to spend more on organic wine for domestic consumption. This result should be periodically reevaluated due to the rapid changes affecting the types of consumption being studied. In contrast, for consumption with friends, it is evident that the organic attribute adds value for the consumer (WTP = 12.70 euros).

7.7. Study 3: direct and incentive measurement of WTP

When confronted with a new food offering, consumers cannot physically test or taste the product presented to them. Consequently, they attempt to anticipate the outcomes of their decision using certain available, relevant and accessible attributes and indicators. Later, after having tasted or consumed the product, they can reassess their evaluation by incorporating sensory information, such as smell, taste, color, etc., into non-sensory attributes and indicators.

The majority of consumers consider wine to be a challenging product to assess and select due to the need to simultaneously consider multiple attributes and indicators. These attributes can be categorized into two groups: intrinsic and extrinsic attributes. When a wine has never been consumed before, the evaluation of expected quality at the point of sale is primarily based on the perception of non-sensory indicators (e.g. AOC, label, price, chateau, etc.), as there is no sensory interaction with the product, unlike fresh products, such as grapes. Therefore, it is pertinent to inquire whether the presence of the organic attribute influences the perception of wine quality and value, both before and after tasting the product.

As a result, the goal of a new study, building upon previous research, was to evaluate the impact of the presence of the organic attribute on hedonic evaluation and to assess its influence on the overall consumer perception, measured through perceived quality and the WTP, both before and after tasting. This was achieved using a pseudo-bidding incentive mechanism based on the Becker–DeGroot–Marschak (BDM) mode.

7.7.1. *Wine selection*

Considering the objective of this study, the first distinctive attribute selected was the type of wine cultivation (conventional or organic). The others were chosen through consultations with both experienced and inexperienced experts and consumers, as well as based on empirical studies related to wine consumption. Ultimately, the study included the following attributes: wine color (red), the Saint-Émilion appellation, which is well-known and recognized by consumers, and the vintage year 2012. To select wines, we collaborated with the "viti-œnology" team at Bordeaux Science Agro, consisting of oenologists and sensory analysis experts. Judges chose one organic wine and one conventional wine. The organic wine was a certified and labeled (AB and EU) bottle of Saint-Émilion 2012, sold at a supermarket for 10.86 euros. The conventional wine was a Puisseguin Saint-Émilion 2012 vintage, available at the supermarket for 9.90 euros.

7.7.2. *Protocol*

The BDM mechanism is designed to encourage individuals to disclose their preferences, as measured by WTP. A participant's motivation to honestly reveal their true WTP is linked to the real consequences of either responding honestly or not, and disclosing their preferences. Specifically, the consequences of each possible behavior fell into two categories: the positive consequence of providing an authentic response was the opportunity to purchase the favored product at a good price, while conversely, the negative consequence of dishonesty was the certainty of not benefiting from a potential good deal.

The experiment consisted of (i) assessing the two wines based solely on non-sensory information and (ii) sensory evaluation of the two wines:

– Initial stage (non-sensory information): each participant assesses a wine using two informational levels. Level 1 includes color + AOC + vintage, while level 2 comprises color + AOC + vintage + organic attribute. Based on the provided information, participants were required to evaluate various wine characteristics, including taste, quality and WTP. To facilitate this evaluation, we used eight-point single-item measurement scales (ranging from 0 to 7).

– Second stage (tasting): the tasting process involved a simultaneous triangular presentation. First, participants were presented with a glass containing conventional wine and two glasses containing the same organic wine. Second, only one of the organic wine samples displayed the organic label. Third, participants were instructed to evaluate the color, aroma and taste of the wine. These evaluations were measured using eight-point hedonic scales (ranging from 0 to 7). To eliminate any order effect, individuals were given the option to choose the order in which they tasted the three samples.

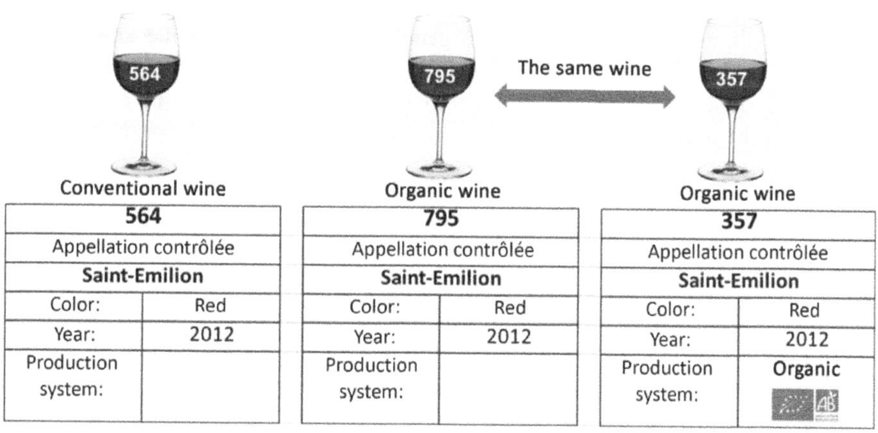

Figure 7.1. *Information and coding of wines tasted. For a color version of this figure, see www.iste.co.uk/cherlet/wine2.zip*

All sessions were conducted under consistent conditions, accommodating a maximum of 16 participants, in a tasting room with individual cabins where the temperature was controlled. The wines were served in standard clear INAO glasses (20ml) at a temperature of 18°C. Each glass was labeled with three-digit random numbers (see Figure 7.1). Additionally, all participants were provided with a glass of water to cleanse their palate between tastings.

7.7.3. *Organic attribute and sensory evaluation*

When presented with the three wine samples, participants were unable to detect and mention significant differences in color. Therefore, it can be concluded that the presence of the organic attribute does not affect the evaluation of the wine's color. Similarly, while the aroma of conventional wine received slightly higher ratings, statistical analysis did not reveal any significant differences. Hence, it appears that the presence of the organic attribute does not influence the evaluation of aroma.

Prior to tasting, consumers perceived organic wines as having a better taste compared to "conventional" wines. This aligns with the findings of certain studies on organic foods. However, this observation contradicts other studies that focus more specifically on wine; these studies often associate a negative image with organic wines (Remaud and Sirieix 2012).

After tasting, conventional wine received the highest rating (5.0), organic wine without indication scored 4.6, and organic wine with the organic label scored 4.7. This result aligns with the perceived organoleptic superiority of the conventional wine chosen for the experiment. Regarding taste, consumers evaluate the two samples of organic wine (with and without the organic label) equivalently. The presence of the organic attribute does not impact the taste evaluation of the wine.

7.7.4. *Evaluation of perceived quality*

Before tasting, the presence of the organic label appears to have improved the perceived and expected quality of the wine for participants. After tasting, similar to the strict taste evaluation, conventional wine received the highest rating (5.0 points), followed by organic wine with the organic label (4.9 points), and in third place, organic wine without the organic indication (4.6 points). These results suggest that consumers a priori expect higher quality from wine with the organic label. The organic attribute positively influences the expected quality of a wine.

After the tasting, results show a significant difference in the perceived quality between organic wine without a label and conventional wine. Conversely, the difference in perceived quality between organic wine displaying the organic attribute and conventional wine is statistically insignificant. Therefore, the presence of the organic attribute seems to significantly and positively influence the quality experienced by the consumer.

7.7.5. *WTP evaluation before and after tasting*

The results that emerged before tasting, based solely on non-sensory information, yielded an average WTP of 10.10 euros for a red wine from Saint-Émilion of 2012 and an average WTP of 11.40 euros for a red wine from Saint-Émilion of 2012 and from organic farming, a difference of almost 13%. After tasting, the WTP for conventional wine was 9.40 euros, 9.20 euros in the case of organic wine displaying the organic attribute, and 8.20 euros for the same organic wine without the organic attribute.

Before tasting, the observed difference between the WTP for organic wines mentioning their organic characteristic (or not) was 12.2%. The percentage difference between organic wine without the organic attribute and conventional wine was 14.6%, while the difference with organic wine displaying the organic attribute was only 2.1%. The presence of the organic attribute, ceteris paribus, can be thus said to increase the purchase value of a wine expressed by its WTP. In the cases considered, it would appear that the WTP adds value to the product.

After tasting, when comparing the WTP of the two samples of organic wine, with and without the indication of the organic attribute, there was a difference of approximately 1 euro (or 12.2%). Since it was the same wine, the difference can be explained only by the presence of the information that the organic label carries. When comparing organic wine without the explicit organic attribute and conventional wine, the difference between the two WTPs was more significant, approximately 1.20 euros. However, when comparing organic wine displaying the organic attribute and conventional wine, the difference between their WTPs was not significant. Therefore, the presence of the organic attribute can be said to increase the value of a wine that would otherwise be less well-valued. The presence of the organic attribute seems to influence the perception of value before and after tasting.

7.8. Conclusion

The research presented in this chapter was driven by the ambition to explore the determinants of the value of organic wine. To achieve this, the study adopted the WTP concept as a measure of behavioral intention. The organic attribute was considered to offer both individual and collective benefits. Methodologically, we chose to employ various WTP measurements (indirect and incentive-based) to assess and elucidate the purchasing and consumption value of the organic attribute in the context of wine.

In doing so, this study sought to at least partially escape the limitations of conventional, non-incentive measurements of behavioral intention, which are typically carried out using scales. At the same time, it also aimed to complement econometric approaches of WTP with the use of a theoretical foundation for the research, considering the history of behavioral intention.

This study incorporates the moral factor into the analysis of organic food consumption by examining universal personal values according to Schwartz et al. (2012). In this context, the research initially demonstrated the relationship between personal values and attitudes towards purchasing and being willing to pay more for organic wine. Therefore, personal values were considered as antecedents of the consumption behavior of organic food, particularly wine. The study confirms the influence of these personal values on the consumption behavior of organic wine in terms of attitudes and evaluation of the offer, especially on the WTP of an organic wine.

Perhaps even more fundamentally, this research has revealed consumer characteristics that can influence the WTP for an organic wine. Overall, the results provide insights into the importance and value of the organic attribute in the wine industry. In fact, it can be concluded that when the intrinsic quality of the wine is similar, the presence of the organic attribute enables producers to justify a higher price for their product.

7.9. References

Achtziger, A. and Gollwitzer, P.M. (2008). Motivation and volition in the course of action. In *Motivation and Action*, Heckhausen, J. and Heckhausen, H. (eds). Cambridge University Press, Cambridge, 272–295.

AgenceBio (2021). Baromètre de consommation et de perception des produits biologiques en France. Report.

Ajzen, I. and Dasgupta, N. (2015). Explicit and implicit beliefs, attitudes, and intentions. The role of conscious and unconscious processes in human behavior. In *The Sense of Agency*, Haggard, P. and Eitam, B. (eds). Oxford University Press, Oxford, 115–144.

Beck, A., Kahl, J., Liebl, B. (2012). *Analysis of the Current State of Knowledge of the Processing and Quality of Organic Food, and of Consumer Protection*. FiBL Deutschland, Frankfurt.

Breidert, C. (2006). *Estimation of Willingness-to-Pay: Theory, Measurement, Application.* Deutscher Universitätsverlag, Wiesbaden.

Fotopoulos, C., Krystallis, A., Anastasios, P. (2011). Portrait value questionnaire's (PVQ) usefulness in explaining quality food-related consumer behavior. *British Food Journal*, 113(2), 248–279.

FranceAgriMer (2019). Ventes et achats de vins tranquilles – Bilan 2018. Report.

FranceAgriMer (2020). Les chiffre-clés de la filière Viti-Vinicole 2009–2019. Report, Montreuil.

Hashibe, M., Shield, K., Zato, W., Vecchia, C., Boffetta, P. (2016). Cancer incidence and mortality attributable to alcohol. *International Journal of Cancer*, 1387, 1380–1387.

Hill, C. and Laplanche, A. (2010). La consommation d'alcool est trop élevée en France. *Presse Medicale*, 39, 7–8.

Jonis, M., Soltz, H., Schmid, O., Hofmann, U., Trioli, G. (2008). Analysis of organic wine market needs. *IFOAM Organic World Congress*, 16–20.

Joshi, Y. and Rahman, Z. (2015). Factors affecting green purchase behaviour and future research directions. *International Strategic Management Review*, 3(1–2), 128–143.

Le Gall-Ely, M. (2009). Définition, mesure et déterminants du consentement à payer du consommateur : synthèse critique et voies de recherche. *Recherche et Applications En Marketing*, 24(2), 91–113.

Remaud, H. and Sirieix, L. (2012). Les vins éco-labellisés face aux vins conventionnels en France et en Australie. Quelle perception des consommateurs et quelles implications marketing ? *Décisions Marketing*, 67, 23–36.

Sagoff, M. (2008). *The Economy of the Earth: Philosophy, Law, and the Environment*, 2nd edition. Cambridge University Press, Cambridge.

Schwartz, S.H., Cieciuch, J., Vecchione, M., Davidov, E., Fischer, R., Beierlein, C., Ramos, A., Verkasalo, M., Lönnqvist, J.E., Demirutku, K. et al. (2012). Refining the theory of basic individual values. *Journal of Personality and Social Psychology*, 103(4), 663–688.

Thøgersen, J. (2011). Green shopping: For selfish reasons or the common good? *American Behavioral Scientist*, 55(8), 1052–1076.

Willer, H., Schlatter, B., Trávnícek, J., Kemper, L., Julia, L. (eds) (2020). *The World of Organic Agriculture. Statistics & Emerging Trends 2020.* Research Institute of Organic Agriculture (FiBL), Frick, and IFOAM – Organics International, Bonn.

8

International Comparison of Adverts for a French Champagne: A Necessary Legal and Cultural Adaptation

8.1. Introduction

In France, due to stringent regulatory guidelines (Évin law), wine brands have to employ creative strategies to promote their products (Duquesnois and Cheriet 2021). However, when they aim to market their products internationally, they still encounter various constraints, albeit less severe ones. On the one hand, several countries have laws similar to the Évin law[1], which impose certain restrictions. On the other hand, it is essential to acknowledge the cultural aspect, as brands must consider it to avoid creating awkward or unsuitable communications.

The necessity to adapt to both targeted and already penetrated markets contribute to a level of complexity in formulating communication strategies for an industry that continues not to invest much in terms of time, interest and money in marketing endeavors (Duquesnois and Cheriet 2021).

However, as the authors point out, one particular wine stands out in this regard: Champagne. Unlike other players in the industry, producers of this globally renowned sparkling wine invest significant time and resources in communication. This is likely due to the vastness of the markets involved and the distinctiveness of

---

Chapter written by Emmanuelle SAUVAGE and Eliane KARSAKLIAN.

For a color version of all the figures in this chapter, see www.iste.co.uk/cheriet/wine2.zip.

1 An example is the Italian IAP (*istituto di autodisciplina pubblicitaria*) which does not specifically focus on the prohibition of wine consumption, but sanctions messages deemed inappropriate, for example, those with sexist or racist connotations.

this prestigious wine, making global communication essential. Champagne is the most internationally recognized French wine, and the mere mention of the word conjures up immediate imagery worldwide, although with some variations. It carries a certain magic, evoking associations with exceptional occasions, celebrations and sometimes solemnity, often symbolizing luxury and always prestige. This wine has, owing to its success, evolved into a standalone brand. Designated as a UNESCO World Heritage Site since 2015, it ranks among France's most dynamic vineyards, achieving record export sales: in 2021, it reached 3.6 billion euros (excluding taxes from Champagne), and during the same year, international shipments amounted to 179.5 million bottles[2]. Hence, it is no surprise that champagne brands implement dynamic national and international communication strategies.

Among all of the producers, one brand stood out to the authors as particularly intriguing: Moët & Chandon (referred to as M&C in this chapter), owing to its global leadership in champagne[3] sales. With distribution in 140 countries, M&C boasts a rich history[4] spanning over three centuries, demonstrating a notable longevity in the industry and remarkable growth. This growth unequivocally signifies its ability to adapt to diverse market contexts and navigate the fluctuations encountered over time[5].

How does this brand manage to adapt to both cultural and specific national laws?

Let us remember that in the realm of international marketing, the prevailing trend for several decades has been for brands to embrace a global communication strategy, promoting a single, supposedly universally applicable message. Fewer brands maintain country-specific campaigns, and it is relatively rare for various national agencies to develop campaigns independently, even apart from the brand's own marketing-communication department. However, M&C does not follow this conventional path; they employ a hybrid approach, adapting their strategy based on the specific target markets.

---

2 Source: https://maisons-champagne.

3 A total of 64.7 million bottles sold in 2019, 95% of which were exported, and 66.8% in 2021. The brand is present in more than 140 countries, with in order the United States (33%), Europe, excluding France (18%), and Asia, excluding Japan (24%) (source: 2019 and July 2022 LVMH financial documents).

4 The brand, founded in 1743 by Claude Moët (Moët & Chandon in 1833), belongs to the LVMH group, created in 1987.

5 "Demand from key markets, with the US and China leading the way, has shown robust growth. Champagne volumes increased by 27% compared to 2020 and by 3% compared to 2019 […] with the reopening of restaurants and the resurgence of tourism" (source: LVMH via Xerfi, July 2022).

This particular aspect warrants a comprehensive and comparative analysis of two campaigns aimed at foreign markets that the brand has implemented in recent years. The first campaign is specifically tailored for the Chinese market, probably thanks to its current and future economic[6] significance. The second campaign, which excludes China and France[7], is a global initiative intended for the rest of the world. The purpose of this double analysis is to assess the contrasting portrayals of champagne and the cultural imagery evoked in each campaign, highlighting the underlying cultural distinctions. Employing a semiotic approach, drawing from the works of Floch (2003, 2010) and Barthes (2004), enables a nuanced examination of the underlying cultural elements shaping the interpretations and subtle references used by the advertiser.

8.2. Semiotic analysis of adverts

Since we live in a world of signs and meanings, adverts are powerful entities in this regard. The study of rhetoric, and particularly semiotics, encompasses both the text (syntactic and paradigmatic analyses) and the context (especially cultural), elucidating the creation of meaningful discourse and the subsequent local and potentially international reception it garners.

International advertising is a world filled with rich symbols and intriguing connotations (Debray 1994), prompting us to unravel and decode messages conveyed through both verbal and nonverbal means. In fact, a study involving consumers from 10 different countries for nonverbal advertising revealed nearly as many diverse interpretations (Schembri and Karsaklian 2014). Our analysis is grounded in complex and multimodal systems (verbal, visual, spatial, sensory), thus centering on a three-tiered semiotic analysis:

– epistemological: aesthetics, culture, communication, narratology, through the use of a descriptive grid;

– conceptual: image, identity, brand, relationship, mental images, social representations, mediation, imaginary, discourse, through cultural analysis;

---

6 Especially since March 2013, following the enactment of new regulations by the Chinese General Administration of Quality Control, which stipulated that only the appellation "champagne" is authorized for wines produced in Champagne. The simultaneous adoption of a Chinese transliteration ("香槟") by the State Intellectual Property Office of the People's Republic of China (SIPO) solidified this legally (source: *Le monde du droit*, September 6, 2013). In 2014, growth resumed, driven by a change in target audience (youth) due to strict anti-corruption laws that have prohibited government officials from consuming alcohol since 2012 (source: viti.com, 2015).

7 For convenience, in the rest of this text, we will refer to it as an international campaign.

– thematic: iconography, environment, commitment, through the context created by and for the visuals.

After conducting a thorough analysis of images using a descriptive framework[8], it is then possible to proceed with an examination in the context of the advertising structure, as defined by Serre-Floersheim (1993) (refer to Box 8.1).

– A hook.

– One or more visuals: photo, drawing, photomontage.

– An editorial block.

– A base line.

– A signature: logo (pictogram or ideogram, acronym) and slogan.

Box 8.1. *Advert structure elements as defined by Serre-Floersheim*

Furthermore, the author proposed other elements to analyze adverts that will be elaborated in a comparative perspective in section 8.5.

8.3. One champagne, two campaigns

M&C launched two international campaigns on different occasions, each targeting separate markets. We chose to conduct a thorough study of these campaigns using the semiotic approach outlined below. On the one hand, there is a campaign specifically developed for the Chinese market. This choice is likely driven not only by the lucrative commercial opportunities in the current Chinese market but also by the relatively recent liberalization of the wine and spirits industry in China[9]. It underscores the absolute strategic necessity of being present in the market during its early development years. On the other hand, there is an international campaign designed for markets worldwide, excluding China and France.

M&C is particularly interesting, primarily due to the combination of different communication choices, local campaign versus international campaigns. This subject has long been resolved, as evidenced by the extensive body of work dedicated to it. We understand that advertisers must trade-off between the potential benefits of locally tailored and differentiated communication and the substantial

8 Grid based on E. Karsaklian's doctoral work: "Measuring the memorization of advertising messages by children: the influence of the school system as a cultural product, in an international approach using four methods" (HEC Doctorate, HEC Group 1995).

9 Study of the Business France office in Shanghai, 2021.

costs associated with it, both in terms of time and budget (coordination, agency selection, translation issues, photo selection, verification, etc.). The decision to have a dedicated campaign for China and a unified campaign for the rest of the world provides an opportunity to analyze the sources of effectiveness for each respective campaign.

8.3.1. *Campaign for China (2022)*

The advert campaign for China[10] showcases a rosé champagne. The inclusion of a couple highlights the romance associated with how the Chinese perceive France and Paris, often referred to as the "City of Romance". In fact, each year, millions of Chinese couples choose to get married in France. In 2021, there were 7.6 million, in 2020, 8.1 million, and in 2019, 9.3 million such weddings, with the peak occurring in 2013 when approximately 13.5 million marriages took place[11].

Figure 8.1. *Chinese couple celebrating their wedding in Paris (photo credit © Aurélie G., TesBellesPhotos)*

Pink serves to evoke both romance and the color of the product. This product is enjoyed both indoors and outdoors, in a setting reminiscent of a garden. The final poster concludes by featuring a bottle of champagne alongside two brimming glasses and some fruit, with the couple notably absent. In the background, glimpses of the garden and a pristine white curtain from the house are visible, mirroring the immaculate white tablecloth on the table. The image exudes an ethereal quality, creating a sensation of almost heavenly lightness.

10 Readers can access these images by clicking on the following links: https://weibo.com/u/1912281322?refer_flag=1005050003_ and https://docs.google.com/document/d/1VMRFaIZCSck3p3jtjdlWQf5R36q2Q_IHTXxyBR47AjE/edit.

11 Courrier International; https://www.courrierinternational.com/article/demographie-la-chine-enregistre-son-plus-faible-nombre-de-mariages-depuis-1986.

In China, love is envisioned as eternal, and the visuals emphasize the enduring nature of couples over time, always accompanied by champagne. Due to its French origins, champagne is seen not merely as a delicious or refreshing beverage, but as a symbol of love and romance[12]. The text narrates romantic scenarios in the couple's life, where champagne plays an integral role, firmly associating M&C champagne with the concept of love.

The individuals featured in the adverts are not well-known, indicating that this champagne can resonate with everyday people. M&C has also employed celebrities in adverts other than this romantic campaign. However, in contrast to the global advertising approach, the brand has opted for a national celebrity: Jing Boran[13] is the brand ambassador in China.

When it comes to the significance of love, each person may have a different response. M&C holds the belief that every love is distinctive and warrants special attention and care. It is about the joy of the initial encounter and the enduring excitement of being together over time. Love is ever-evolving; it is about sharing the pleasures of life and narrating them thoughtfully to one another. Every moment of love is precious and deserving of celebration.

This campaign was deliberately launched on May 20, 2022, a date that, when written in Chinese, appears as 2022/05/20, or 520, and phonetically resembles "I love you" in Chinese. It is a clever play on words, capitalizing on the phonetic similarity of the two phrases[14]. Furthermore, the presence of so many "2" digits in a single date is uncommon, and the sum of these digits, which is the number 8, is considered auspicious in Chinese culture[15] as it symbolizes luck and success. This number is featured in the advertising tagline that continues with, "The Moët &

12 For the linguistic processing of Chinese, the authors received assistance from a Chinese national, whom they would like to thank here.

13 See: https://fr.wikipedia.org/wiki/Jing_Boran.

14 "I love you" (我爱你) is pronounced as *wǒ ài nǐ* and is homophonic with "520", which is pronounced *wǔ èr líng* (Northern accent, Mandarin) or *wǔ èr níng* (Southern accent). Since the 1980s in China, the economic development (light industry) of the southeast has been faster than that of the north. The south therefore has a greater influence on the market. When it comes to "520" people have become used to the pronunciation of the south.

15 In China, the number 8 is associated with luck (捌, pinyin ba1), as it sounds like the word for "prosperity" (發, pinyin fa1). Synonymous with wealth, success and social significance, it is therefore favored by the Chinese, because it holds importance in Chinese traditions as in modern life (e.g. 8 is preferred in license plates, the address of a building, telephone numbers and even the rounding of promotional promotions, which will end in 98 and not 99 as in the West) (source: voyages-chine.com).

Chandon champagne campaign tells the story of love with you and bears witness to everything that matters to you":

> The moments shared with you are truly dazzling: your shyness, your smile, and the times we reminisce about, I want to capture them all. As the bubbles rise in our wine glasses, let's secretly cherish the love that can't be concealed; feel the bittersweet essence of our love and sense its irreplaceable nature. Sitting in a sunlit room, toasting to us, it's incredibly delightful to have you by my side. Under the soft glow of dimmed lights, enveloped in the sweet aroma of Moët & Chandon champagne, surrounded by love like enchanting bubbles, in this moment, we belong exclusively to each other.

Six visuals follow:

– The third year after we met, our love could not be hidden.

– To love and be loved simultaneously is a joyful experience.

– Staying together for a long time with excitement is better than the joy of a sudden reunion.

– 2,000 days of life with you, you are there when I am awake and when I dream.

– There are always surprising encounters, like when I meet you.

– Love, not just in 520, every moment of love is worth remembering.

8.3.2. *International campaign (excluding China and France) (2011)*

M&C's dynamic communication over several decades highlights its constant adaptability to different situations and contexts. Early on, the brand expressed a desire to align with the communication standards of luxury brands in the realms of perfumery, high jewelry and leather goods. Continuing its commitment to evolving its strategy, the brand made a significant move in 2009 by shifting from the world of fashion, which was previously lucrative but becoming overcrowded, to the world of cinema. In this transition, they selected a young American actress, Scarlett Johansson, as their ambassador. Associating an alcohol brand with an ambassador is not always straightforward, especially for the ambassador who must be willing to have their image associated with an alcoholic product. In this case, the brand made a safe choice by selecting a rather uncontroversial Hollywood actress with no known scandals to date. In 2011, they followed this decision with a campaign featuring the young actress captured through the lens of the renowned British photographer Tim Walker. Although initially intended for a global audience, M&C had to forgo broadcasting the campaign in France due to legal restrictions (see Box 8.2).

It was an ad campaign consisting of four visuals[16] and a series of photos of Scarlett Johansson as Marilyn Monroe.

Figure 8.2. *Iconic photo of Marilyn used as inspiration for the Scarlett campaign (photo credit © Wikimedia commons)*

Figure 8.3. *Example of material and color used in the Scarlett campaign (photo credit © www.veaul.com)*

Scarlett Johansson strikes sensual poses in all images from the campaign. She is seen wearing either an haute couture dress, a sky blue drape that unveils her bare legs and bust, or white veils.

16 Readers can access the images by clicking on the link: https://www.journal-du-design.fr/pop-culture/illustration-publicite-moet-chandon-et-scarlett-johansson-14058/#top.

"The law does not exclude editorial articles from the definition of advertisements, nor does it require that such articles impose a burden on the brand they feature. It is sufficient for the article to promote an alcoholic beverage beyond the limits set by Article L3323-4 of the public health code, especially by associating alcohol with a festive context or featuring influential personalities, for it to be considered in violation of legal provisions. The photographs depicting a young, famous and cheerful actress with a bottle of champagne, along with a flattering commentary [...] clearly exceed the permissible statements".

Box 8.2. *Extract from the judgment of the 5th Chamber of the TGI Paris of March 21, 2013, relating to the prohibition of M&C's "Scarlett Johansson" campaign*

The actress's relationship with the product changes depending on the image, while keeping the product and its brand clearly visible. The actress interacts with the product in all of the images without actually being in a situation of consuming the product. Scarlett Johansson is in different positions in each image: she holds a bottle while climbing stairs next to empty pyramid-shaped glasses from the floor to the ceiling; she sits on a chair next to bottles on the floor without touching them, the bottle is at her feet on the opposite side of a sofa in a garden; and finally, in a sensual pose on a chair, she holds a bottle by the neck, letting it touch the floor as a drunk person would. A dozen empty bottles are arranged on the floor around this chair, resembling the aftermath of a party, all set in the interior decor of a castle.

a) b)

Figure 8.4. *a) Cascade of champagne glasses. b) Queen's Chamber, Palace of Versailles (photo credits © Wikimedia, CC BY-SA 3.0)*

The message of sensuality (Scarlett Johansson was voted the sexiest woman in the world in 2006 and 2013[17]) and French sophistication is clearly conveyed through this set of images, thanks to the reference to "historical heritage" through the furniture, the location and the architecture that exude French prestige and chic. The combined image of luxury and simplicity is evoked, particularly by the designer haute couture-style clothing (rolled-up sheet), the grandeur of the location and the abundance of bottles. The advert is signed "Moët & Chandon Champagne" and completed by a line that reads "Scarlett Johansson. Maison Moët & Chandon. Épernay. France".

M&C's President explained the main reason why the brand retained an American brand ambassador[18]:

> Scarlett Johansson was a clear choice as an ambassador, because she, like Moët, has a story to tell, a magical story that captivates people. She is a genuine and refreshing icon, spontaneous, generous, seductive, and fully embraces life both on and off screen.

Scarlett Johansson is globally renowned for her roles in films that have been widely distributed. She stands as one of the most iconic contemporary actresses in the film industry. Her ability to portray characters reminiscent of the glamor and romance of the 1950s–1960s, such as Grace Kelly and Marilyn Monroe (as seen in the visual above), also allows the brand to evoke nostalgia for the times when champagne was considered a sure bet and symbol of glory and success. It has always been and continues to be associated with the imagery of being the drink of choice for stars and the powerful elite of the world.

The interpretation of this ad campaign could (perhaps) delve deeper for fans of the actress who are aware of her love for Paris and France[19]. This insight into the actress's personal life could also contribute to conveying an image of freedom, a certain form of rebellion against the backdrop of the place's classicism and the actress's nonchalant and daring poses; all blended in a subtle mix of simplicity and glamor, exuberance and lightness.

It is worth noting that this is the first time M&C has enlisted a Hollywood actress to represent its brand. The company chose to blend the glamor, elegance, grace and

17 Magazine *Esquire*, October 2013.

18 Frédéric Cumenal, President of Moët & Chandon, spoke about the choice of Scarlett Johansson as the brand ambassador; interview in *Première* dated April 2, 2009.

19 She was married to a Frenchman, Romain Dauriac, and has a French American son as a result.

youth of this international celebrity with the Trianon, the historic venue of Maison Moët[20].

Figure 8.5. *Château Moët et Chandon in Épernay (photo credit © Wikimedia, CC BY-SA 4.0)*

In this blend, we can discern a skillful fusion of tradition and modernity, a common technique in luxury[21] marketing. The brand incorporates elements deeply ingrained in luxury product communication, such as traditional craftsmanship (underscored by the decor), pleasure (noticeable on the actress's face), authenticity (tied to the heritage and historical significance of the locations), allure and prestige. These are the notions that consumers want. Besides the theatrical staging, we can observe the use of classic luxury codes like shine and gilding (Anido-Freire 2014), which brands have embraced since the inception of the industry (Lipovetsky and Roux 2003). Furthermore, these images combine simplicity (in contrast to the opulence of crystal chandelier pendants and gilded door frames, both voyeuristic and demonstrative) with minimalism in terms of objects and language. These two dimensions appear to resonate with champagne consumers (Favier et al. 2019). The actress embodies a modern princess, with the castle setting contributing to this metaphorical representation, indulging in a luxurious lifestyle. Scarlett Johansson, shifting between *ingénue* and *femme fatale*, plays across various roles with blurred

20 Located precisely in Epernay, historical land of the brand.

21 Let us revisit the etymology of the word "luxury": it originates from the Latin word *luxus* (debauchery, excess, splendor), derived from *lucere* (to shine, to be brilliant), which itself comes from *lux* (light, brilliance, illustration, luster, glory, ornament). Luxury is a social practice involving extravagant and unnecessary expenditures, aimed at surrounding oneself with sumptuous refinement or driven by a pure penchant for ostentation, as opposed to factors strictly related to necessity (source: *Petit Robert* dictionary).

boundaries. This can be seen as a subtle invitation to consumers to live a life of opulence. While champagne consumption remains a status experience, consumers can choose to embrace the moment and imagine themselves living in the luxurious world depicted in these photos. Finally, these images also possess an artful quality, turning the product and its consumption into a form of art (Passebois-Ducros et al. 2015). The visuals are crafted like paintings, with aesthetics, composition and photography suggesting works of art. These aspects also evoke a refined and delicate Western way of life, which holds great appeal abroad.

8.4. M&C advert for France taken in the veins of the Évin Law: a way out

Aside from the challenges associated with the various cultural interpretations of adverts targeted at international markets, M&C must also consider the legal framework for its communication within France. In this context, it is noteworthy to showcase an example of M&C advertising, both before and after the legal prohibition decision supported by the Évin law. The French legal framework stands out globally for the level of restrictions imposed on advertisers (Tsikounas 2004).

a) b)

Figure 8.6. *a) Concept of the advert released in 2006 and subsequently banned in France, with the bottle originally featuring the brand specified in the ruling in the box below. b) Concept of the revision implemented in 2013 (source: Pixabay)*

The "La nuit est rose" advert, created in 2006, features a bottle of champagne at the center of a swirling mass of pink petals seemingly bursting from behind the bottle, all set against a black background. A slogan displayed at the top of the poster reads: "The night is rose" (see Figure 8.6), and the brand's logo appears prominently in large gold letters at the bottom of the poster. This advert was banned primarily due to concerns that the slogan "the night is rose" could create an association between the consumption of this rosé champagne and having a euphoric outlook on life, a concept commonly expressed as "voir la vie en rose" (see Box 8.3) in French. This phrase is widely popularized by Edith Piaf's (1946) famous song, known to virtually all French citizens. The judge's use of the term "common language" indicates that, from the judge's perspective, there is no ambiguity regarding the association between these two expressions, and furthermore, this frame of reference is assumed to be universally understood by all French people. The judge implicitly invokes a common framework without explicitly naming it because he considers it self-evident.

The 13th Criminal Division of the Paris Court of Appeal has determined, in a ruling delivered on Wednesday, that the campaign conducted by the wines and spirits group of the French luxury house LVMH was indeed illegal. [...]

In its ruling, the court determined that the slogan "La nuit en rose" ("The night in rose") took this advert beyond the scope of authorized mentions, as it created an association between the consumption of this rosé champagne and having a euphoric approach to life, a common interpretation of "seeing life in pink". The court observed that "to protect public health", the legislator intended to restrict the use of such subjective associations, common in advertising and allow only the most objective references. [...]

In a press release, the ANPAA expressed satisfaction with this ruling, emphasizing that "contrary to the producers' assertions, the Évin law is crystal clear, and their prosecution results from their unwillingness to comprehend and abide by it".

Box 8.3. *Excerpt from the judgment of the 13th Criminal Division of the Paris Court of Appeal regarding the prohibition of the broadcast of M&C's campaign "La nuit est rose"*

In its revised version, the French advertisement promotes rosé champagne through a floral visual, positioning the bottle at the center of a pink flower that conveys the freshness of its wine, the delicacy of its aromas, lightness and fluidity. The background of the poster is also pink. The bottle is the central and sole element of the advertisement. This highlights the fullness of the product, which needs nothing else for its value enhancement. It features a simple, minimalist, monochromatic pink image that emphasizes the dark color of the bottle with its pink label. In addition, the flower cradles the bottle like a case, resembling taffeta, creating an impression of lightness and softness. The various shades of pink subtly allude to the rose and life in pink, evoking a dreamy and perfect life, free from difficulties. The rose symbolizes beauty, *finesse* and delicacy. It conveys a straightforward message that requires little effort in imagination and interpretation. The consumer is encouraged to imagine what this image evokes for them and establish personal associations. This is a form of indirect communication, akin to an "instruction to imagine", as highlighted by some researchers (Bréhonnet and Lagier 2019).

By subtle circumvention of the prohibition, we ultimately find the initial elements present in the first advert, by evocation, by suggestion, a metonymic form of the flower: we no longer have a cloud of petals, but the flower in its entirety.

8.5. M&C, an exemplary use of semiotics in marketing

From a visual standpoint, the French advert consists of a single visual, whereas the campaigns for China and international markets include multiple visuals, six for the former and four plus one for the latter. It is worth noting the complete absence of a baseline in the French and international adverts, in contrast to the Chinese campaign. The presence of the signature and logo is evident in all visuals. A more in-depth analysis, as suggested by Serre-Floersheim (1993), should address the following questions:

– What are the key strengths of the featured product? The shared strength across these adverts is the prominent presence of the product and the brand's reputation. The common theme among them is the concept of luxury.

– What needs does it address? In addition to the physiological needs that any beverage would fulfill, the needs highlighted in these advertisements are rather subtle. Unlike adverts for other product categories, these do not revolve around a problem-solving approach. Instead, they imply enjoyable situations that are fitting for a luxury champagne. Champagne symbolizes success for those who can afford to celebrate personal and professional achievements with it (hedonic dimension). It represents both self-satisfaction and self-expression (symbolic dimension).

– What advantage will the advertiser leverage? The primary advantage here is the product and the brand. They are sufficient to become the primary asset in the lives of individuals who consume it, in various situations.

– What is the promise? The promise is full of beauty, sensuality and romance.

The author also asserted that an advertising message must be monosemic, meaning that all the elements present in the visuals are complementary and converge to convey a single message. Furthermore, the author described the various codes as analytical elements in adverts (see Box 8.4).

1) Graphic code: all of the adverts examined in this study are printed visuals disseminated through media outlets like magazines.

2) Typographic code: there is no text in the French and international adverts. The Chinese advert uses simple, classic ideographs.

3) Linguistic code: valid only for the Chinese advert.

4) Social code (social class, emotions): the social code is defined by luxury that targets the upper social classes of the population in all three cases.

5) Kinetic (gestural) code: limited to the static interaction of the actors with the product.

6) Proxemic code (use of space): the use of space in the adverts is divided between individuals and the product for Chinese and global adverts, while it is exclusively represented by the bottle within the flower for the French advert, with a complete absence of individuals.

7) Chromatic code (warm and cool colors): the colors used are rather subtle and cool to emphasize the product. Pink dominates in the French ad. It is the color associated with painters of the Rococo era. François Boucher and Jean-Honoré Fragonard painted Madame de Pompadour in pink attire, which became a symbol of fashion, extending to Sèvres porcelain. Pink historically represented women of character. This fashion trend even reached the United States when Marilyn Monroe adopted it as her favorite color at the time. Today, pink symbolizes femininity, delicacy and love, a symbolism undoubtedly linked to the flower, rose.

8) Referential code (culture): as pointed out by Iribarne (2002), when, unknowingly, words and images are not assigned the same meanings, it can lead to misunderstandings. Cultural diversity necessitates adaptations for understanding messages from different cultures. While companies often lean towards anchoring their advertising messages in so-called universal values, such as love, family, sensuality, among others, they overlook the fact that the expressions of these values vary across different cultural contexts (d'Iribarne 2009).

Box 8.4. *Codes according to Serre-Floersheim*

Consistently, the global advert in which Scarlett Johansson undeniably resembles Marilyn Monroe is predominantly in shades of pink. For the other adverts in the same campaign, bluish-green and white dominate the actress's clothing. Green symbolizes nature, hope, eye and mind relaxation. It signifies life (*viridis*), represented by living beings like plants (Scarlett Johansson is surrounded by vine leaves). Green heralds the arrival of spring when all plants start to adorn themselves in young green leaves. Today, "being green" signifies protecting nature or affiliating with the ecological political party, the Greens. While white symbolizes purity in some cultures, it represents mourning in others and even combines elements of both, as seen in China. The use of white conveys lightness, neutrality and discretion, allowing extravagance to be associated with other colors.

Figure 8.7. *Champagne vineyard (photo credit © Wikimedia, CC BY-SA 3.0)*

In the historical backdrop where the actress is situated, we encounter shades of beige, brown and gold. Beige derives its name from the natural color of wool provided by sheep and was coined by the French. It is not an expressive color, as it is considered neutral, but it is valued for its ambiguity, allowing it to harmonize well with other colors. Brown symbolizes the earth that gave birth to human beings, sustains them and receives them after death when their bodies return to the soil. However, brown is also associated with dirt and mud. This color gained greater appreciation thanks to Impressionist painters who depicted vibrant outdoor nature where brown is ubiquitous. It is worth noting that the grape clusters used to make champagne originate from the earth, and historic paintings in places like the Lascaux caves in France and the Hands cave in Argentina were all created using earth pigments. Finally, the color gold signifies rarity, thus sophistication and wealth. It falls under the category of heraldic[22] colors and continues to shape our social symbolism today (Divard and Urien 2001).

22 This classification was born in the West in the 12th century to be able to identify combatants on the battlefield.

Figure 8.8. *Golden taffeta (photo credit © Craftine)*

As a result, European royalty was significant consumers of products and clothing made of gold fabric, as worn by Kings Henry VIII of England and Francis I of France. Gold is also a legendary color, referencing the myth of the Greek god Midas, who turned everything he touched into gold.

Pink and white dominate the Chinese advertising campaign, although green and brown are also present. While colors may carry different cultural meanings, those used by M&C exude universality. The neutral tones of beige aim to avoid overshadowing the product, which should be perceived as the central element in all advertisements. The presence of plants and the colors green and brown highlights the natural aspect of the product derived from the earth and vineyards. The green bottle further emphasizes this natural connection while using gold to signify sophistication. Finally, pink, closely associated with rosé champagne, conveys the delicacy of the grape varieties used in making this refreshing wine.

It is not possible to associate a color, which is undoubtedly a form of discourse or a way of acting, with a value(s) while ignoring the differences between cultures (Gouret et al. 2013; Celhay et al. 2020). In international communication, only a very vague vocabulary that allows for multiple interpretations can be used, so that everyone, regardless of their culture, can relate to it. However, when faced with the same stimulus, each person will bring their own interpretation (Celhay et al. 2017).

An essential aspect of intercultural relations lies in how, in the absence of a common context for interpretation, individuals influenced by distinct cultures may interpret symbols differently (Geertz 1973). For instance, the definition of romanticism is not universal. Scarlett Johansson's sensuality could be viewed as romantic or not by individuals from other cultures, and the Chinese couple may not convey love as ardently as a French or American couple would. Regarding the French advert, only those familiar with the creative references behind the expression "la vie en rose" could interpret it beyond the apparent freshness of the product associated with the freshness and elegance of the flower.

According to Barthes (1964), the meaning of an image in advertising is intentional: certain attributes of the product form the a priori signifieds of the advertising message, and these signifieds must be transmitted as clearly as possible. The three adverts analyzed in this chapter share the common theme of luxury. They convey a suggestive message rather than a descriptive one, using cultural stereotypes and imitation of behaviors to a certain extent, but not excessively.

A globally renowned brand can afford to adapt its messages worldwide because, as Umberto Eco (1972) suggests, the effectiveness of advertising relies on consumers' a priori acceptance of the message. According to the author, we accept advertising persuasions because we are convinced by reasoning that is already familiar to us, and it typically concerns things we already desire rather than something advertising tries to make us desire. Advertisers leverage what is already known and use established solutions. In this context, advertising rhetoric aims to determine the extent of its reach. Consequently, our analysis reveals the use of well-known codes in all three adverts: "la vie en rose" in France, the sensuality of Marilyn Monroe worldwide, and French romanticism in China. This approach undoubtedly predisposes a global audience to consume the product.

We were able to gather spontaneous reactions from a small international sample of 30 people[23], including Chinese and American people, without any claims of scientific rigor. What emerged is that the locations used in the international campaigns are not recognized as French or even French-style, and the reference to Marilyn Monroe does not appear to enhance its value.

For the Chinese campaign, it seems quite natural to celebrate the date of 520, even though traditionally Valentine's Day is celebrated in February. Regarding the French advert, based on the responses, it appears that having a brand ambassador or text is not necessary because the French are already familiar with the brand, as per the respondents. In addition, the visual has at times been interpreted as having a sexual connotation. This highlights the potential gap between the presumed intentions of the advertiser and the actual reception by consumers. It underscores the importance of paying attention to the process of conveying meaning and the genuine receptivity of a message, both in terms of nature and intensity.

23 Exercise carried out in August 2022 in an academic setting.

8.6. Conclusion

Everything has to change for nothing to change[24].

The analyses produced made it possible to enrich or consolidate the reading grid of M&C's cultural reasons for making the necessary adaptations.

Furthermore, they have highlighted the inability to transfer images without altering their original meaning from one cultural context to another. This raises questions about the universality of the receptiveness of a conveyed visual message and its consistent acceptance on a global scale, whether in this particular case or in any international brand deployment scenario. It is important to note that this does not imply a systematic need to modify content, images or text, but rather emphasizes that advertisers should be aware of potential variations in how the message is interpreted due to its cultural underpinnings, which may not always be consciously acknowledged by the parties involved. In some cases, there may be a form of denial regarding the intended meaning, akin to the challenges of translation in linguistic contexts. These variations can potentially impact the core objective of any brand, which is to foster a positive consumer attitude. Therefore, the decision between choosing a local or global communication campaign should be part of a broader evaluation of desired objectives within the overall brand strategy, rather than solely a budgetary consideration.

In the realm of international advertising, there are instances where it is advantageous to contemplate what can be termed as a semiotic translation, akin to linguistic translation. This involves making the reflexive act of transposing the meaning of an image as apparent as it is for a written text when you intend to convey it to someone who speaks a different language than the sender. The consideration, or lack thereof, of the various connotations of an original message, extending beyond its denotation, should be subject to informed discussion or, at the very least, scrutiny.

8.7. References

Anido-Freire, N. (2014). When luxury advertising adds the identitary values of luxury: A semiotic analysis. *Journal of Business Research*, 67, 2666–2675.

Barthes, R. (1964). Rhétorique de l'image. *Communications*, 4, 41–42.

Bréhonnet, R. and Lagier, J. (2019). Quand "l'instruction à imaginer" fait évoluer la communication sur les vins. *Décisions marketing*, 1(93), 53–70.

24 *Il Gattopardo*, Giuseppe Tomasi di Lampedusa, Feltrinelli, Milan (1958).

Celhay, F., Masson, J., Garcia, K., Folcher, P., Cohen, J. (2017). Design graphique du packaging et innovation : une étude comparative des codes visuels des vins de Bordeaux et de la Barossa Valley. *Recherche et applications en marketing*, 32(2), 48–75.

Celhay, F., Cheng, P., Masson, J., Li, W. (2020). Package graphic design and communication across cultures: An investigation of Chinese consumers' interpretation of imported wine labels. *International Journal of Research in Marketing*, 37, 108–128.

Debray, R. (1994). *Manifestes mediologiques*. Gallimard, Paris.

Divard, R. and Urien, B. (2001). Le consommateur vit dans un monde en couleurs. *Recherche et applications en marketing*, 16(1), 14–16.

Duquesnois, F. and Cheriet, F. (2021). Loi Évin, marketing du vin et protectionnisme. *Open Wine Law*, 1(1), 1–17.

Eco, U. (1972). *La structure absente. Introduction à la recherche sémiotique*. Mercure du France, Paris.

Favier, M., Celhay, F., Pantin-Sohier, G. (2019). Is less more or a bore? Package design simplicity and brand perception: An application to Champagne. *Journal of Retailing and Consumer Services*, 46, 11–20.

Floch, J.-M. (2003). *Sémiotique, marketing et communication : sous les signes les stratégies*. Presses Universitaires de France, Paris.

Floch, J.-M. (2010). *Identités visuelles*. Presses Universitaires de France, Paris.

Geertz, C. (1973). *The Interpretation of Cultures*. Basic Books, New York.

Gouret, M., Lo Monaco, G., Urdapilleta, I., Parr, W.-V. (2013). Social representations of wine and culture: A comparison between France and New Zealand. *Food Quality and Preference*, 30, 102–107.

Hmaida, J. and Décaudin, J.-M. (2019). Comment les individus issus de différentes cultures catégorisent les marques ? L'impact de la conception de soi. *Décisions marketing*, 1(93), 93–106.

d'Iribarne, P. (2002). *Cultures et mondialisation*. Le Seuil, Paris.

d'Iribarne, P. (2009). La force des cultures. *Le débat*, 5(157), 111–123.

Lipovetsky, G. and Roux, E. (2003). *Le luxe éternel : de l'âge du sacré au temps des marques*. Gallimard, Paris.

Passebois-Ducros, J., Trinquecoste, J.-F., Pichon, F. (2015). Stratégies d'artification dans le domaine du luxe. Le cas des vins de prestige. *Décisions marketing*, 80, 109–124.

Schembri, S. and Karsaklian, E. (2014). Who sees what? One print advertisement and a dual semiotic analysis. *International Journal of Marketing Semiotics*, 2, 63–80.

Serre-Floersheim, D. (1993). *Quand les images vous prennent au mot ou comment décrypter les images*. Éditions d'Organisation, Paris.

St Clair, K. (2016). *The Secret Lives of Color*. Penguin Books, London.

Tsikounas, M. (2004). Quand l'alcool fait sa pub. Les publicités en faveur de l'alcool dans la presse française, de la loi Roussel à la loi Évin (1873–1998). *Le temps des médias*, 1(2), 99–114.

PART 3

Collective Strategy and Terroir

9

Wine Terroirs and Collective Images of Wines: The Beaujolais Rosé Case, Failure or Potential?

9.1. Introduction

Adapting vineyards' offerings has always been a crucial aspect of their development throughout history. It is essential for them to stay aligned with the changing expectations of their contemporaries in various markets (Brouard 2021).

However, the offerings from several French vineyards experienced a positioning crisis that started in the 2000s. This crisis began with a decrease in both volume and value of sales, followed by a decline in production. Without repositioning, this type of crisis ultimately leads to vineyard removals and the disappearance of various entities, including estates, cooperatives and businesses.

Therefore, the production of Beaujolais increased from 1,111,444 hl in 2005 to 588,729 hl in 2015[1], due to the crisis of its main offer, that of en primeur wines, a trend from the 1960s to the 1990s.

This mispositioning of the Beaujolais offer resulted in media bashing, labor union tensions, and, most importantly, a historic wine crisis. To address this situation, the French government appointed its Inspector General of Agriculture, Georges-Pierre Malpel, in the mid-2010s. His report[2] diagnosed a structural crisis,

---

Chapter written by Jérémy ARNAUD.

1 French newspaper *Les Échos*, dated November 17, 2016.

2 Report No. 15058, Georges-Pierre Malpel, Inspector General of Agriculture, December 2015.

proposed long-term repositioning strategies, recommended a new governance approach, but did not include a specific marketing strategy. The marketing strategy was developed in the second half of 2016 by the consulting firm Terroir Manager, which is our team, in response to a request from Inter Beaujolais, the interprofessional organization representing all 12 AOPs in the Beaujolais vineyard, including Beaujolais, Beaujolais-Villages and various crus (Brouilly, Chénas, Chiroubles, Côtes de Brouilly, Fleurie, Juliénas, Morgon, Régnié and Saint-Amour).

Approved in early 2017, this strategy was titled "Beaujolais Nouvelle Génération" (Arnaud and Brouard 2016). Its goal was to foster the growth and enhancement of a "new generation" of Beaujolais wines, encompassing all AOPs, while preserving and even amplifying the vineyard's DNA of drinkability and conviviality. This is achieved through an upscale approach to production and diversification into white and rosé varieties (Brouard 2021).

The incorporation of rosé into this strategy is of particular interest, considering that this type of wine has become quite indispensable, especially in France. However, it still accounts for less than 3% of Beaujolais volumes. Only the Beaujolais and Beaujolais-Villages AOCs are authorized to produce it, not the crus. From 2017 to 2021, their average production in this category amounted to 14,666 hl, with Beaujolais AOC alone contributing more than 80% of this production[3].

In addition, concerning the current and potential position of rosé in the Beaujolais offering, we will sequentially examine, in broad strokes, the viewpoints of the vineyard, wine consumers in France, and Terroir Manager, the consultant.

9.2. The vineyard's feedback

9.2.1. *Producers*

Terroir Manager first posed five questions about the current, potential and desired positioning of Beaujolais in the rosé category to a group of around 10 producers from private cellars and cooperatives, considered representative by the ODG Beaujolais and Beaujolais-Villages.

In response to the first question, "Why does Beaujolais produce so little rosé?", the primary reason cited is the lack of a collective effort in this regard. More specifically, the explanation points to a deficiency in the vineyard in terms of expertise, technical and financial resources, volume purchased by the trade,

3 Source: Inter Beaujolais.

interprofessional promotion and communication, competitiveness, and even legitimacy regarding this particular wine offering (see Table 9.1).

Explanation levels	Iconic verbatim	Identical verbatim
Expertise	"It's not in our culture."	"No collective skills and shared visions on this matter," "lack of a chromatic index or a standard technical process," "oenologists' expertise is inadequate in this regard," "everyone contributes too little," "requirement for training and information to approach this type of wine seriously and produce it effectively," "numerous gaps," "rosé is made as an afterthought," "monoculture: single grape variety, color, and nearly the same wine."
Technical and financial resources	"We don't have what it takes."	"It's challenging with gamay," "we struggle to make it age without pinot noir," "we lack the equipment, except for cooperatives and some estates," "there are technical and financial barriers because it requires substantial resources."
Trade	"The demand is either insufficient or too late."	"What we produce is what we sell," "if the demand was there, we would produce," "to produce it, we must have the orders, because there is a risk of mis-sale," "there is no trade to buy it from us," "there is a trade that innovates and a trade that only responds to calls for tenders; however, in our region, there are not enough innovative traders, moreover, they ask us for rosés at the last moment! It's the same problem in white," "we have to anticipate, impossible at the last minute."
Marketing	"There is no dedicated collective-communication."	"No interprofessional collective support," "when we sell it, we are all alone," "if no marketing, no demand and therefore no production," "Beaujolais is not known for producing rosé," "today, when we sell Beaujolais rosé, we primarily sell rosé."
Market	"Competition is strong."	"Languedoc is half as expensive."

Table 9.1. *Roundtable of representative producers (source: Terroir Manager)*

In response to the second question, "What interest would Beaujolais have in producing more rosé?", multiple reasons are provided. Some are financial, as rosé is a wine for quick consumption and can generate cash flow, similar to early wines. Others are related to the decline in red wine consumption, especially in France, and the need to compensate for it; this is considering that rosé can also be made with gamay. In addition, the interest in rosé for developing new markets and seizing new opportunities is mentioned, as there is now a growing demand for this type of wine, both in France and abroad, including strategic markets such as Great Britain and the United States, which are important for Beaujolais. Ultimately, regardless of the reasons cited, the producers present agree that producing 50,000 hl of Beaujolais and Beaujolais-Villages rosé would be a relevant medium-term goal.

In response to the third question, "What positioning do you recommend for Beaujolais rosés?," a consensus emerges regarding their profile. They should be dry, without sweetness, offering freshness and liveliness similar to Sancerre, and have a relatively light color, avoiding excessive pressing, akin to the style found in Provence. In essence, these rosés should "stay awake," as one producer put it, and be suitable for consumption at the table. There is also agreement on positioning these rosés in the mid-range, priced at 7 to 8 euros, including tax per 75 cl bottle at the winery for the consumer. This positions them "between the entry-level wines of Languedoc and the high-end rosés of Côtes de Provence". Overall, it appears strategic to present Beaujolais rosés with a distinct style to create a unique market segment, as it is recognized that "we will never be like Provence"; therefore, "we must stay true to our identity", as emphasized by some producers. Regarding commercial development, the suggestion is to initially prioritize short distribution channels and then transition to longer distribution channels using established wholesalers and traders. There is also the issue of a festive event, "to be found", to promote the rosés of Beaujolais.

In response to the fourth question, "How should Beaujolais organize itself to successfully develop a rosé offering?", several key points emerge. First, there is a need to establish a dynamic collaboration between cooperatives and traders, similar to the model seen in Prosecco production. This entails traders reassuring cooperatives by informing them, no later than the end of August, about their intentions regarding rosé purchases, specifying details such as volume, quality and price. One producer emphasizes that it is crucial to acknowledge Beaujolais' culture of turnover, indicating a preference for having money in the bank rather than excess stock. In addition, it is essential to create a cooperative environment among producers, facilitated by economic interest groups (EIGs), and offer support services from cooperatives to winemakers who may not have the necessary equipment. In general, this organization should focus on improving skills and nurturing a pool of proficient rosé producers. Finally, the responsibility lies with Inter Beaujolais to formulate a collective marketing strategy, to allocate a dedicated interprofessional budget, and, consequently, initiate a chain reaction to expedite market penetration. This strategic approach is expected to accelerate market entry.

Regarding the fifth question, which pertains to "what communication do you advocate for Beaujolais rosés?", a debate ensues. Some producers believe that it is not advisable to strongly link the rosé offering with Beaujolais. They argue that "People want rosé, not necessarily specifically Beaujolais rosé." Instead, they suggest that the IGP Rhodanian Counties might be a better fit for this type of wine. Another viewpoint is the need to establish and promote a separate collective brand for these wines.

Some regret that rosés have been authorized for Beaujolais primeurs, "because it's not the right choice". Conversely, some producers consider it strategic to promote the name Beaujolais, and therefore Beaujolais or Beaujolais-Villages rosés, in order to gain identity, while they believe "we lose it with Coteaux Bourguignons", a regional appellation created in 2011 that can be claimed by Beaujolais winemakers located in the Rhône and Saône-et-Loire. Finally, there are producers who recommend promoting gamay rosé, because "abroad, they ask for the grape variety", and some of them specify that it is better to write "gamay noir" than "gamay" on the label.

9.2.2. *Wine research center*

After consulting with this group of producers, Terroir Manager proceeded to schedule a meeting with Bertrand Chatelet, the director of Sicarex, a viticultural research center in Beaujolais, to gain insights into the potential of this vineyard for rosé production.

On the oenological front, there are certain constraints to consider. "First of all, the gamay grape variety is not very aromatic," explains Bertrand Chatelet, "which implies harvesting at the right maturity. However, climate change should work in its favor." This aspect is particularly crucial because "Beaujolais rosés are exclusively produced from gamay grapes (mandatory single variety)," and therefore, "there are no other red grape varieties available to create blends with more distinct aromatic profiles, as seen in Provence, for instance." Nonetheless, he highlights some positive aspects. For instance, "this grape variety allows for adaptation to acidity and color preferences in demand." In general, he acknowledges that "there are still some aspects to be validated in terms of technical processes. Malo, non-malo[4], and what about the choice of yeasts? One thing is certain: oenological *process* developed for Provence rosés can be easily adapted for use in Beaujolais; it's just a matter of fine-tuning. We should conduct experiments and develop tools to provide reassurance to winemakers." Regarding rosé, Chatelet notes,

4 Malo is the abbreviation of malolactic fermentation, which allows for the transformation of malic acid into lactic acid via anaerobic bacteria called lactic acid bacteria.

> We have a database with approximately 350 analyses, but it remains largely untapped. We could, in conclusion, leverage the existing results and establish a group of producers and traders focusing on rosé, consisting of leaders as well as volunteers. The goal would be to validate consistent technical approaches in line with the new interprofessional strategy and conduct an initial assessment based on the 2019 vintage.

However, Bertrand Chatelet believed that "the true barriers to the development of rosé production in Beaujolais" are more related to socio-economic concerns than technical ones. In other words, it is primarily about convincing traders about the volume and remuneration levels. According to him, it is essential to "determine the grape destination very early, down to the plot level, and therefore implement contractual agreements for outlets to encourage winemakers to produce rosé".

9.2.3. *Interprofessional organization*

After meeting with the oenologist director of Sicarex, Terroir Manager got in touch with Anthony Collet, who was the marketing manager at Inter Beaujolais, to assess the sales of Beaujolais rosés in French retail stores.

At a commercial level, Anthony Collet described a positive trend driven by the new vintage wines.

> From 2011 to 2018, Beaujolais rosés increased by 4.3% in volume[5], but it is important to note that this growth is largely attributed to the new Beaujolais rosés[6]. Sales of Beaujolais and Beaujolais-Villages rosés excluding the new vintage releases represented significant volumes until 2012, serving as a crucial diversification avenue at that time. However, they sharply declined afterward, almost becoming negligible by 2018[7]. In general, he adds, retailers recognize the need for diversification to present the Beaujolais category differently and leverage the renewed interest in this vineyard. Nevertheless, they are currently primarily interested in the new rosés because selling Beaujolais and Beaujolais-Villages rosés, excluding new vintage releases, is challenging in the face of offerings from the Loire, Provence or Languedoc.

5 318,668 liters in 2018 compared to 305,567 liters in 2011 (source IRI).

6 285,124 liters in 2018 compared to 171,992 liters in 2011 (source IRI).

7 33,544 liters in 2018 compared to 133,575 liters in 2011 (source IRI).

While the developments are indeed promising, there are still several challenges to address. First, “some retailers believe that there is room for qualitative improvements” in Beaujolais rosé. In addition,

> they consider that the sources of supply need significant expansion since the volumes currently offered are far too low for permanent listings. The few available references are either regional or found at wine fairs. Lastly, they suggest that Beaujolais should organize in-store promotions to raise awareness about its rosés and boost sales. They also encourage the exploration of new formats or innovative packaging.

9.2.4. *Marketers*

Several leading players in Beaujolais were finally interviewed by Terroir Manager to find out their vision and their dynamics on subject of rosé.

In terms of quality and style, Thierry Letortu, the General Manager of Alliance Vignerons Bourgogne Beaujolais, believed that “today, harvesting gamay grapes at optimal ripeness allows us to meet the expectations of rosé consumers who seek fruitiness, freshness, and moderate alcohol content”. On the contrary, Frédéric Berne, a winemaker in Lantignié, shared that after producing “an initial rosé with fruity and delicate characteristics”, labeled as “gamay noir”, he planned to “elevate the quality by crafting rosés from different terroirs, including pink granite, blue stones, siliceous soils, and more”.

In the marketing aspect, François Roth, the director of the Agamy cooperative winery, emphasized that the rosé market he is targeting “is highly influenced by trends, which necessitates adaptation. At Agamy, he specified that they primarily sold their Beaujolais rosés in large retail chains, catering to buyers who currently preferred pale, slightly tangy, fresh, full-bodied, fruity rosés with citrus notes. We are currently introducing a three-liter cubic Bag-in-Box (BIB) called ‘Carrément Rosé’”.

On the commercial front, Thierry Letortu shared: “In 2018, we produced 1% of our Beaujolais in rosé. This will increase to 5% in 2020, and our medium-term goal (5–7 years) is 10%. From a commercial perspective, we have already created and tested cuvées in the French market, including the Ce n’est pas la mer à boire brand for mass distribution, and the Premier Bain and Rose Nacarat brands for traditional channels. In 2020, we will introduce new brands to target specific wine merchants, wine bars, and bistronomic restaurants, both in France and internationally.” François Roth also shared his insights: “In 2019, our sales of Beaujolais nouveau rosé increased both in France and internationally, and we anticipate a further 50% growth in 2020. We are actively expanding our presence in the retail sector, and we will now consistently offer Beaujolais rosé to both traditional and specialized channels, with a particular focus on the restaurant industry."

9.3. Consumers' feedback

9.3.1. *Survey type*

After obtaining the vineyard's feedback, Terroir Manager conducted a survey with a representative group of French consumers through Ipsos, one of the world's largest market research firms.

In terms of methodology, this survey was conducted from April 12th to April 19th, 2019, among a random sample of 1,041 French individuals aged 18–75 years.

In line with Terroir Manager's instructions, it explored the potential consumption of Beaujolais rosés, making a distinction between:

– Beaujolais nouveau rosés, to be enjoyed starting from the third Thursday of November, which is the release date of Beaujolais nouveau. These are therefore early wines intended to be consumed young, i.e. in the first months following their harvest.

– Beaujolais rosé, to be enjoyed year-round and available from January 15th, the release date of red Beaujolais.

9.3.2. *Results*

First, this survey offered contextual information.

It confirmed the now unavoidable place of rosé in wine consumption in France: three quarters of French people consume wine and 48% of them rosé (51% in the 25–45 age group). It also verified the leadership of Provence rosés: 32% of rosé consumers buy them (38% of over 45s).

It also revealed that rosé is more favored among the French compared to Beaujolais. 53% of them enjoy[8] rosé wines, while only 35% have an affinity for Beaujolais wines, and 25% appreciate the Beaujolais Nouveau festival on the third Thursday of November. In other words, 31% do not like rosés, 39% do not like Beaujolais and 45% do not like the Beaujolais Primeurs festival. The preference for

8 Percentage including those who "like" and "like a lot".

Beaujolais wines is even lower among individuals aged 25–45 years (26% like it and 49% do not).

Another highly intriguing result was that Beaujolais consumers had less affection for rosé wines (47% of them) compared to consumers of red (67%), white (71%) and rosé wines (84%) in general.

Second, this survey made it possible to take stock of intentions to consume:

– they are higher for Beaujolais rosés (29%) than for Beaujolais nouveaux rosés (24%);

– they are higher, regardless of the type of Beaujolais rosé, new or otherwise, among people who already consume rosés (46% intentional for Beaujolais rosés, 37% for Beaujolais nouveaux rosés), and less strong among those who already consume Beaujolais (32% for Beaujolais rosés and 27% for Beaujolais nouveaux rosés);

– finally, a large majority of intentionalists plan to consume rosés from Beaujolais, regardless of the type, in addition to (and not in place of) their current consumption of rosé wines or Beaujolais.

Third, this survey confirmed positioning keys. It indicated that consumers of wines in general, Beaujolais, and even more so rosés, widely agreed[9] to say the following about the two types of Beaujolais rosé:

– that they must be creative in terms of bottle and label (even more so for Beaujolais nouveaux rosés);

– that they give a modern image of this vineyard;

– that they must seek to compete with the best rosés in terms of quality.

Regarding Beaujolais nouveaux rosés specifically, consumers of Beaujolais wine and even more those of rosés, largely agreed to say:

– that they are in the spirit of the traditional Beaujolais Nouveau festival (end of November);

– that they make you want to participate in the Beaujolais Nouveau festival;

– furthermore, only a minority believe that Beaujolais nouveau rosés are exclusively consumed as an aperitif, and just a slight majority say that these rosés are well-suited for the month of November.

9 Percentage including those who “agree” and “strongly agree”.

Regarding Beaujolais rosés, in particular, wine consumers, Beaujolais enthusiasts and especially rosé lovers largely agreed to say:

– that they must be distinguished from other rosés by their taste;

– that they should be offered more for sale in stores;

– that they are consumed during meals;

– that can be associated with all types of dishes.

Finally, and always in terms of positioning, this study delved deeper into certain marketing aspects:

– When it comes to wine consumers in general, the buying potential is at its peak at 5 euros including tax (25%) for both types of rosé. However, it is higher for Beaujolais rosé after 8 euros, especially between 10 and 15 euros including tax.

– Concerning rosé consumers, the buying potential for Beaujolais rosés is higher than for Beaujolais nouveau rosés from around 10 euros.

– The color of Beaujolais rosé is associated with a slightly darker hue than that of Beaujolais nouveau rosés.

– The preferred bottle shapes for Beaujolais rosés are more traditional compared to Beaujolais nouveau rosés.

Figure 9.1. *Summary of the IPSOS study (source: Terroir Manager)*

9.4. Terroir Manager's feedback

9.4.1. *A strategic offering*

Since the 2000s, rosé has become a standalone market, and its consumption has witnessed spectacular growth in certain markets (Brouard 2019). This is particularly true in France, where in 2020, rosé accounted for nearly 35% of the volumes purchased for home consumption, compared to less than 10% in the 1990s[10].

10 Source: Iri.

In France, the growth of rosé wine is primarily attributed to a shift in purchasing behavior from red to rosé wines. In addition, there has been an increase in exclusive buyers who exclusively purchase rosé, meaning they only buy this type of wine in the market (source: Secodip, current Kantar).

The growth of rosé abroad is particularly linked to certain major markets, such as Great Britain and especially the United States, but other important potential markets are now forecast, such as China.

Rosé wine consumption is becoming more year-round, although it still experiences peaks and even records during the summer. As a result, there is now a category referred to as "winter rosé" both in France and the United States (winter and inside conviviality vs. summer and outside conviviality).

Its positioning has significantly increased in added value, influenced in particular by the "elegant rosés of Provence" whose style (color, taste, packaging) serves as a significant source of inspiration for the competing offerings in the southern France region. This trend is also seen in the broader context of a growing appreciation for "thirst-quenching wines" regardless of their color.

It must be noted that the Beaujolais vineyard has stayed away from this growing market since the 2000s, the very years when the new Beaujolais entered into crisis.

Today, it is likely that the primary rosé markets have reached maturity, allowing for a diversification of offerings in this category. However, this diversification must maintain a high level of quality to compete effectively and provide a unique selling proposition compared to existing products (a new trend of alternative rosés).

In this new context, the "Beaujolais Nouvelle Génération" strategy aims to expand its rosé offerings. This aligns with the findings of the Ipsos study conducted in spring 2019, which appears to confirm the recommendation made by Terroir in January 2017. According to this recommendation, rosé could serve as a strategic addition to the "Beaujolais Nouvelle Génération."

Indeed, with the right positioning, a rosé offering could help avoid the decline in red wine consumption, modernize the brand image, attract new buyers and generate additional sales. This collective momentum towards rosé primarily concerns the range of Beaujolais en primeur wines, which have a festive character and still account for one-third of its total production. A revival of Beaujolais Nouveau is possible, but it should be of a new generation, partially rosé. According to Terroir Manager, the key is for the new generation to embrace the trendy rosé culture, adopt its codes and marketing mix, rather than the other way around. Once integrated, the third Thursday of November event remains its primary asset.

In general, to effectively serve as a strategic complement to the range, the Beaujolais rosé offering must meet two key criteria. First, it should align with market expectations in terms of volume and quality, which entails dedicating a larger and more consistent portion of the harvests, despite their susceptibility to climate-related fluctuations. Second, the rosé offering should maintain alignment with the overall ambitions and portfolio of the vineyard, ensuring that it does not inadvertently revive negative perceptions and bashing associated with the Beaujolais nouveau rouge that started from the 2000s.

9.4.2. *An offering strategy*

In its January 2017 recommendations, Terroir Manager believed that a more significant diversification in white, as well as in rosé, can and should contribute to the emergence of a "Beaujolais New Generation" offering, recognizing that three categories of wine were distinguished.

First, there were the "Beaujolais de fête" through the volumes of Beaujolais nouveau and Beaujolais-Villages nouveau. These now accounted for only a third of the total production of this vineyard, averaging around 200,000 hl, compared to 50% at their commercial peak. They constituted the traditional foundation of the Beaujolais offering. However, Terroir Manager recommended a collective strategy to revalorize this volume of "thirst-quenching wines" by continuing to capitalize on the global event dedicated to them every third Thursday of November. As a result, the development of the offer in dry festive rosés was proposed up to 15 euros including tax per 75 cl bottle ex-winery (range creation). An offering of sparkling wines was also considered relevant after discussions and tastings on the subject at Sicarex, but the appellation regulations for Beaujolais and Beaujolais-Villages do not allow this type of wine, whether it is nouveau or not. White wine is also prohibited as a nouveau wine, but it is not strategically positioned as a "Festival Beaujolais" due to its very low marketable volume and its potential as a terroir wine, rather than a simple thirst-quencher.

Second, there are the "Beaujolais of character", especially through the volumes claimed in Beaujolais-Villages, and also, and most importantly, in cru wines. With an annual average of 400,000 hl and a bistronomic spirit, they form the new foundation of the Beaujolais offering. Terroir Manager described them as "thirst-quenching terroir wines", and they are positioned between 7 and 15 euros including tax, with the need to primarily be between 10 and 15 euros by the end of the 2020s and never below 10 euros for cru wines. The "Beaujolais Nouvelle Génération" strategy encourages diversification of this offering with characterful white and dry rosé wines, bearing in mind that this is legally possible in Beaujolais and Beaujolais-Villages, but not in cru wines.

Third, there are the "Exceptional Beaujolais" produced in limited volumes primarily in the crus as well as in Beaujolais and Beaujolais-Villages, where applicable. With just a few thousand hectoliters produced, they represent a pioneering category of new-generation terroir wines, emphasizing micro-terroirs (including specific named locations) and priced starting from 15 euros including tax. The new strategy also recommended developing an offering of exceptional dry white wines within this category, considering that, legally, this could only apply to Beaujolais and Beaujolais-Villages. Terroir Manager did not propose diversifying into rosé within this category.

In this triptych offering, two types of new-generation Beaujolais rosé are positioned, each with a role to play as a range complement.

Regarding "Beaujolais rosés de fête" (festive Beaujolais rosés), primarily priced between 7 and 10 euros including tax, the challenge lies in aligning them with the trends in the world of rosé wines, rather than keeping them rooted in the traditions of Beaujolais from the past. This transformation aims to attract a new customer base, particularly among those under 45 years of age who are regular rosé consumers. By incorporating these rosés more prominently and aggressively into the offering on the third Thursday of November, it could rejuvenate and reinvigorate this event, which has seen declining interest since the 2000s. However, it is worth noting that even during this decline, 20 million bottles of "Nouveau by Beaujolais" red wines continue to be purchased in France and worldwide. This approach would also allow the Beaujolais vineyards to take the lead in the niche of "winter rosés"[11] or "festive winter rosés", which have gained popularity in the northern hemisphere. These winter rosés have unique characteristics – they are the first rosés of the year, facing competition mainly from rosés of the previous year and other rosé en primeurs. They are associated with the conviviality of winter, including specific ambiance, meal types and activities. To succeed, the festive winter rosés from Beaujolais need to differentiate themselves from the current "new" red and rosé wines, which often tend to be quite conventional. They should position themselves as a fresh offering with a strong emphasis on quality, offering a "clear winter color" slightly less clear than summer rosés, and distinctive and original packaging (with imaginative branding and visuals on the commercial facing and clear appellation labeling, such as Beaujolais Nouveau or Beaujolais-Villages Nouveau on the back label). This should all be combined with an inherent "joyful fruitiness" on the organoleptic level. In this endeavor, the role of the wine trade is strategically significant.

Regarding the "Beaujolais rosés de caractère", priced from 10 to 15 euros including tax, it is recommended to maintain or adopt Beaujolais-specific codes for

11 Opinion Way survey (2018): 74% of French wine drinkers consume rosé in winter, 83% are 18–49 year olds.

them and offer them as a complementary range to a clientele typically over 45 years old, who are already familiar with or favorable to this vineyard. To achieve this, the Beaujolais vineyard should leverage its individual statutory brands, especially those of the estates, positioned at a minimum of 10 euros, typically crafted in red, sometimes in white, and now also available in rosé, with the same packaging and at the same price point. In essence, these niche rosés, crafted by winemakers and released as early as mid-January, targeted at prescriptive distribution channels, can contribute to boosting the sales volume of "Beaujolais of character" in general, which is the new strategic foundation of the vineyard. These "Beaujolais bistro-style rosés" (in reference to bistronomy – encompassing the circuit, ambiance and dishes) distinguish themselves from other rosés both in appearance (featuring a rich color, darker than that of Provençal rosés) and in flavor (characterized by a vibrant fruitiness). By positioning themselves in this manner, these "table rosés" assert a Beaujolais identity or authenticity (highlighting the winemaker and origin on the main label), but they must also adhere to the current market constraints of rosé wines, including price and taste preferences.

9.5. Conclusion

Figure 9.2. *Beaujolais Pavilion at Wine Paris 2020 (photo credit © Terroir Manager). For a color version of this figure, see www.iste.co.uk/cheriet/wine2.zip*

In February 2020, following the "Beaujolais Nouvelle Génération" strategy, the interprofessional organization, for the first time, created a promotional showcase for Beaujolais rosés at the Beaujolais Pavilion at Wine Paris.

However, for this new rosé offering strategy to truly thrive and succeed, it must be embraced by the vineyards, cooperatives and wine merchants themselves. As a result, Inter Beaujolais, in partnership with the Rhône Chamber of Agriculture and the Auvergne-Rhône-Alpes Region, has introduced a business support program called "Beaujo Coaching" since 2022.

9.6. References

Arnaud, J. and Brouard, J. (2016). Stratégie Beaujolais Nouvelle Génération. Terroir Manager Report for Inter Beaujolais.

Boudon, L. and Lacroix, P.-A. (2019). Évaluation du potentiel de consommation des rosés du Beaujolais. IPSOS Report for Terroir Manager.

Brouard, J. (2019). Le rosé, les leçons d'un succès. Chaire Unesco Culture et Traditions du Vin [Online]. Available at: https://chaireunesco-vinetculture.u-bourgogne.fr/ressources-chaire/annales/conferences/202-les-conferences-2018-2019.html.

Brouard, J. (2021). Le Nouveau Beaujolais est arrivé. Un vignoble nouveau en phase avec son époque. *10ème édition Journées Pontus de Tyard*, 54–66.

10

Collective Strategies and Positioning: The Case of the Loire Volcanique Initiative

10.1. Introduction

Several collective initiatives are emerging in the French wine industry to highlight common characteristics and consolidate often limited resources. These efforts are characterized by various forms of governance, diverse construction processes, which are often poorly controlled, and distinct objectives (e.g. promotion of a region through local markets, launch of a collective brand, pooling of production or marketing resources, collaborative efforts to access land, the consolidation of communication tools, the development of commercial events).

In addition to their diversity, these collective entrepreneurial initiatives (Cook and Plunkett 2006) are marked by a high degree of complexity and dynamic construction processes (Renting et al. 2011). They are heavily influenced by national and regional institutional contexts, as well as their relationships with other stakeholders in the relevant region (Amblard et al. 2020). In practice, these initiatives can either be promoted or hindered by specific public actions, as observed in the case of certain collective entrepreneurial endeavors focused on varietal wines in Languedoc-Roussillon (Blancanaux 2019).

Academic research on collective entrepreneurial initiatives in agricultural sectors has received limited interest. These studies emphasize the importance of a "comprehensive and dynamic analysis" of the construction processes (Renting et al. 2011), the identification of barriers and success factors for farmers' collective business endeavors (Knickel et al. 2008), the consideration of other "external factors" related to tourism and regional development (Dana et al. 2014),

Chapter written by Foued CHERIET, Hervé HANNIN, Clément ARNAL and Fanny LEPAGE.

performance analysis (Loubaresse and Pestre 2016), the roles of leadership (Yan and Sorenson 2003), connections with other stakeholders in the region (Amblard et al. 2020) and governance choices (Schieb-Bienfait et al. 2020).

This chapter aims to illustrate the steps, strengths and challenges in constructing a collective strategy through the presentation of a research action approach, conducted between 2016 and 2018, with vineyards in the Massif Central region of France (Côtes d'Auvergne, Saint-Pourçain, Côtes du Forez and Côte Roannaise). These small vineyards are characterized by unique identities, different and sometimes conflicting expectations, and distinct project management approaches. Using a methodological research-accompaniment approach (Bréchet et al. 2014), the objective is to demonstrate how the process of establishing a positioning for these four vineyards reconciled the specificities of individual structures with the requirements of a constructed collective strategy.

This research is grounded in a theoretical approach to collective strategies, aiming to analyze the development of a collective positioning for the four vineyards as a "collective entrepreneurial project" (Emin and Schieb-Bienfait 2013). This contribution demonstrates that, despite significant initial divergences, the involvement of various actors in the co-construction of the collective project enhances its legitimacy and robustness. It also highlights the active and positive role of data collection from stakeholders, prescribers and institutions. In addition, the results shed light on sometimes substantial misunderstandings among the actors themselves (in terms of their expectations and visions) and between them (their knowledge of each other). Similarly, it observes how certain disagreements persist and lead some actors to abandon the project or not become adequately involved in it. Finally, the results reflect the actors' capacity for adaptation and their ability to take ownership of the collective construction's conclusions to continue its development.

Following a brief analysis of "the question of collective positioning(s)" in the wine sector, the four vineyards under study will be introduced, followed by a detailed explanation of the methodological approach. Subsequently, the obtained results will be thoroughly examined and discussed. Finally, the current developments and prospects arising from the new collective positioning of these four vineyards will be outlined.

10.2. Collective positioning in the wine sector

10.2.1. *Positioning: an existential strategic question*

The question of positioning a product line, brand or company is arguably the most critical aspect of any strategic marketing endeavor. It falls at an intermediate level between understanding the internal and external environment through a series of analyses that culminate in a diagnosis and the operationalization phase that involves using the tools of the marketing mix.

Positioning is the strategic translation of a product or brand's identity. It involves making deliberate and strategic choices regarding three distinct components in relation to: (1) our own products or product lines, (2) other products or competitors and (3) our customers. For example, when San Pellegrino, a brand owned by Nestlé, entered the French market in the 1990s, it had to position its mineral water in a way that differentiated it from Badoit and, most importantly, did not place it in the same competitive category as Perrier (also owned by Nestlé) to avoid any "cannibalization". Consequently, San Pellegrino chose not to position itself as a "pleasure" water (a segment occupied by Perrier), but rather as a high-end, "gastronomic" water.

When it comes to competitors, positioning fundamentally aims to highlight distinctive elements (price, quality, brand strength, specialization and other product advantages) or similarities (product range, terroirs, product profiles, etc.). For example, a positioning in relation to competitors can be: "Our product is similar to that of X or Y." Generally, even though there are various dimensions to consider (quality, reputation, production process, range breadth, etc.), marketing practices often emphasize "price positioning" in relation to competitors or simply categorize products as "high-end" or "low-end".

Finally, when it comes to customers, positioning aims to identify a tangible and unique solution to their needs. It involves clarifying one aspect of the product or brand's identity in the customer's mind. This clarification is then translated into practical terms through visual and communication tools (brand image construction, visual codes, labels and slogans) that emphasize various dimensions of the positioning (price, brand or product promise). The question of positioning becomes more complex when a company promotes multiple brands or products. It becomes even more intricate when dealing with collective positioning that combines distinct identities and diverse objectives.

10.2.2. *Approaches to building collective positioning in the wine sector*

10.2.2.1. *Historical sector specificities*

In the initial analysis, the wine sector seems to be in a favorable position for presenting meaningful and robust positioning strategies. It is characterized by multiple "terroirs" whose historical valorization has been closely linked to the concept of "typicity" naturally encompassing the identification of distinctive traits that constitute a positioning strategy. However, this foundational concept of the appellation of origin remains imperfectly defined. Furthermore, the widespread development of collectives and trade associations aimed at creating and managing geographical indications, protected designations of origin (AOP) and protected geographical indications (IGP) was a prevalent trend during the 20th century, eventually encompassing 93% of French wine production. Yet, in reality, this approach, rooted in agronomy, oenology and legal aspects, was primarily led by wine production stakeholders and paid little attention to consumer perception. Consequently, positioning strategies constructed around purportedly distinct features, often asserted solely by producers and insufficiently recognized downstream in the industry, often evolved from "intentional" positions into "unintentional" and poorly controlled ones.

For instance, the division of Languedoc into smaller appellations was a response to a commendable quality-oriented policy and the emphasis on terroir distinctions by producers. While some specialists understood this approach, it still proves challenging to break free from a lingering and persistent perception, at times associated with "*piquette du midi*" (a lesser-quality wine), despite the undeniable quality of its present-day wines. This may be attributed to one of the sector's characteristics, as it has been cautious about adopting marketing strategies, fearing the influence of consumers who may not fully grasp the intricate segmentations in place but are undeniably hard to disregard!

Furthermore, large retailers, known for their extensive wine sections featuring hundreds of options, may have misled various producer associations regarding the market's recognition of the attributes of their terroirs, positioning, reputation and image. In truth, these retailers were primarily focused on retaining their customers, who were long-time, loyal consumers dedicated to their preferred wines. Consequently, in the vast majority of cases, neither producers nor retailers truly addressed the issue of positioning with marketing expertise until the beginning of the 21st century.

10.2.2.2. *Promising effects of relevant positioning*

There are, however, a few rare exceptions to this observation. One such example is champagne, where leading brands are owned by large corporations well-versed in the merits of strategic marketing and the "segmentation–targeting–positioning" triad. In this case, the value of a strong positioning is particularly evident. These wines, more than just fermented grape juice, aromas and carbonation, consistently evoke the terroir of Champagne, celebrations, the accompaniment of joyful occasions and a touch of luxury. Such successes have naturally set a precedent, demonstrating the advantages of a well-crafted, distinctive positioning that is intentional rather than imposed and aligns with the aspirations of a collective; each member who contributed to its development is more inclined to apply it at the level of their own production unit.

10.2.2.3. *Real, persistent difficulties*

It is therefore clear that segmenting the wine landscape into nearly 450 geographical indications, and potentially nearly 1,280 if we include additional designations, sub-regional appellations and crus, will prompt the question of objectifying the distinctive features that form the foundation of these diverse positions. A fortiori, how can these intended positions be aligned with the positions perceived by customers, who may not always be able to discern such a multitude of unique characteristics?

However, this sector presents numerous other sources of complexity, as it requires intersecting these situations with the positioning of "cooperative wines" versus "independent winegrowers", and "organic" wines versus various other signs related to the environment and health (HVE, biodynamics, etc.). This undoubtedly explains the challenge of such an approach in the wine sector. Finally, from a managerial perspective, we can appreciate the difficulty of uniting collectives for this purpose, which is much greater in the case of trade unions and interprofessional organizations than in structured marketing companies with dedicated departments.

10.2.3. *Determinants of collective positioning in the wine sector in France*

There are numerous examples of collective positioning in the wine sector. Building such collective strategies involves multiple levers and connects different actors at various levels (individual, organizational, interorganizational and institutional), often yielding mixed results (Loubaresse and Pestre 2016). These strategies are typically "territorialized" to emphasize the specificities of the terroirs in question.

The primary determinants of collective positioning in the wine industry relate to the diversity of actors and products, the rarity or uniqueness of the terroir, quality attributes, legitimacy based on history or expertise and standards of modernity and creativity. Other determinants may stem from new societal expectations such as environmental commitment, health-related concerns, and the promotion of collective and cooperative efforts.

For instance, the CIVB (Bordeaux wines) emphasized the diversity of their wines a few years ago, positioning them as a range of varied, international options to cater to all consumer needs. Their slogan at the time was "Bordeaux wines, there is so much to discover". In a recent digital campaign, the wines of Alsace embarked on a collective positioning that highlights the eco-responsible commitment of winegrowers and the oenotourism potential of the region.

To promote the diversity of its products, the collective brand Champagne de Vignerons, associated with the Syndicat Général des Vignerons de la Champagne (SGV), recently launched a campaign featuring six visuals of producers, associated with specific product attributes ("woody and mature", "raw and mineral", "golden and fruity", "fresh and iodized", "nature and floral" or "pink and silky"). A few years earlier, the SGV had initiated an unprecedented campaign to "demystify" the product and make it suitable for all seasons through a large-scale campaign with a budget of 4 million euros: "Champagne, suitable for all occasions."

With around 60 AOPs and eight IGPs, Loire wines take pride in their extensive diversity of terroirs, grape varieties and flavors. Yet, they also face the challenge of establishing a "cultural identity" (Peyronnet 2022). As a response, Loire IGP wines initiated a campaign (by the Mineral agency) with the slogan "naturally creative".

The wines of Chablis in Burgundy emphasize the exceptional nature of their product, centered around a grape variety (Chardonnay): "four appellations, one grape", while the wines of Charente showcase the diversity of their product range and the dedication of their winemakers: "600 winemakers, one appellation".

Therefore, determinants of collective positioning may refer to distinctive attributes such as terroir, grape varieties, winemaking methods, as well as various other aspects. These challenges, mentioned earlier, add complexity to the positioning process: which attributes should take precedence? What generic characteristics should be emphasized to showcase the collective richness of such diversity? How can a strong collective identity be reconciled with occasionally opposing individual identities? And how can collective positioning be updated without risking competition with other collective positions in the wine industry?

Over the past decade, there has been a renewal of collective[1] positioning in the Occitan vineyard. While not exhaustive, the following are a few examples from the Occitanie region, which encompasses the Languedoc-Roussillon wine region, showcasing the diverse strategic choices and positioning dimensions within the regional wine industry (see Table 10.1).

1 In this regard, we can refer to AOC Conseil, a company that provides support to numerous collectives and institutions within the Occitanie region's sector.

Collectives	Positioning	Slogan, *baseline*
Languedoc-Roussillon	Diversity and dedication of actors	Shared instinct
Saint-Chinian	Opposition of terroirs on the same appellation	Shale and limestone
Plaimont	Originality/age of the grape varieties of the South-West	*Wines of the South-West with great origins*
Picpoul de Pinet	A wine from the sea and seafood	Its terroir is the sea
Cahors	Cradle of an international grape variety	Cahors Malbec
Castelbarry	Modern products and image, cooperative commitment and quality products	*Artisanal and engaged cooperative*

Table 10.1. *Some examples of collective positioning in the wine sector in Occitanie*

While certain positioning strategies focus on terroirs and pedoclimatic characteristics (e.g. Saint-Chinian), and the uniqueness and historical significance of grape varieties (e.g. Plaimont), others take a consumer-oriented approach (e.g. Picpoul de Pinet). Recently, with the support of the marketing agency Sowine, the wines of the Languedoc-Roussillon region have undertaken a collective image revamp, identifying the sociological profiles of winemakers and the organoleptic characteristics of the wines, emphasizing commitment, attachment to the terroir and strength of character.

The case of Cahors is also interesting because it makes a significant choice in its general positioning (Cahors Malbec), emphasizing an international grape variety, and then extends this concept (Cahors Malbec Lounge, Cahors Malbec Days, etc.) all the way to wine tourism (Destination Cahors Malbec). Finally, the positioning of the Castelbarry winery emphasizes the unique aspects of the cooperative structure (social commitment, craftsmanship and collective work).

10.3. Study background and approach

10.3.1. *Presentation of the vineyards of Massif Central*

10.3.1.1. *Differentiated but close vineyards*

The four vineyards in question are located in the Auvergne-Rhône-Alpes Region in the center of France (see Figure 10.1). The AOP Côtes d'Auvergne is in the Puy-de-Dôme department and Saint-Pourçain is in Allier, both bordering the Allier

River. The Côtes du Forez and Côte Roannaise are situated in the Loire department and are adjacent to the upper course of the river.

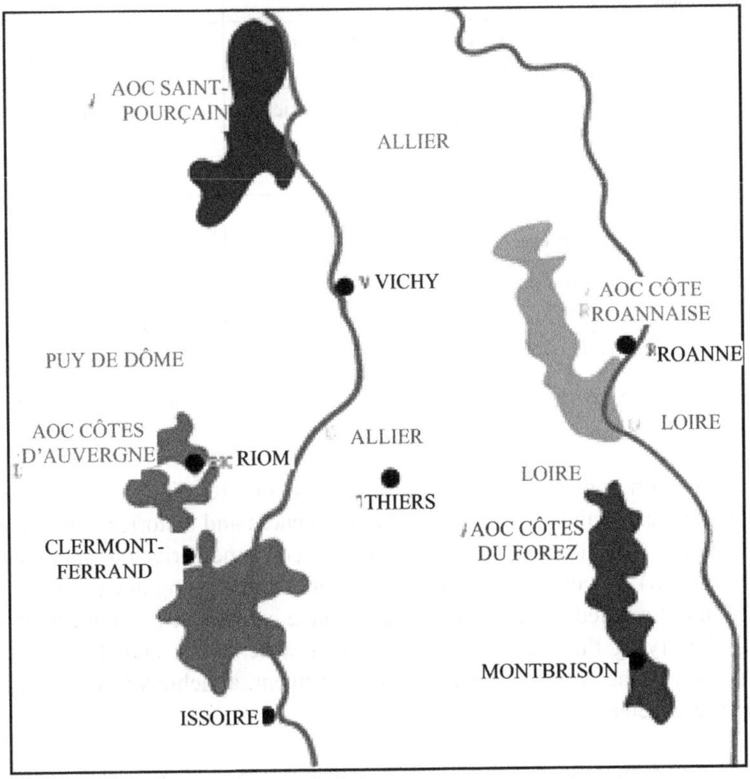

Figure 10.1. *Location map of the vineyards of Massif Central (source: Ici commence la Loire). For a color version of this figure, see www.iste.co.uk/cheriet/wine2.zip*

Each of these vineyards possesses a unique personality and maintains a distinct, albeit often imposed, positioning. They all feature different wine characteristics, organizational structures and appellations (see Table 10.2).

Nevertheless, they do share similar climatic conditions, unique terroirs (granitic, volcanic or clay-sandy) and common dominant grape varieties (gamay, with pinot noir and chardonnay as secondary varieties). Therefore, they have similar viticultural potential. In addition, they all share the common characteristic of having an expansive territory where viticulture is no longer the primary activity, as it is integrated into regions focused on breeding or mixed agriculture.

	Côtes du Forez	Côte Roannaise	Saint-Pourçain	Côtes d'Auvergne
Winegrowing area (hectares)	110	210	591	460
Geographical area (number of municipalities)	17	14	19	53
Number of producers	– Nine independent winemakers – 30 cooperative winemakers	32 independent winemakers	– 18 independent winemakers – 40 cooperative winemakers	– 46 independent winemakers – 50 cooperative winemakers
Production volume (hectoliters)	4,300	8,300	26,000	14,000
SIQO	– AOP Côtes du Forez – Urfé and Rhodanian County IGP	– AOP Côte Roannaise – IGP Urfé and Rhodanian County	– AOP Saint-Pourçain – IGP Loire Valley	– AOP Côtes d'Auvergne – IGP Puy de Dôme

Table 10.2. *Characterization of the four vineyards studied (source: ODG of the four vineyards)*

10.3.1.2. *Shared challenges*

These vineyards predominantly share significant challenges, which are the basis of their merger project. They grapple with a lack of recognition, an absence of understanding of their characteristics and a deficiency in visibility among consumers. They are burdened by an image of "lesser wines", "thirst-quenching wines" that endures and remains deeply ingrained in the perception of these vineyards, both internally among residents and externally among consumers and influencers.

They also face significant regional competition and a lack of representation within the sector's governing institutions (FranceAgriMer, Wine Basin Committee). Both viticultural and regional authorities tend to provide better support to larger regional vineyards (Côtes du Rhône and Beaujolais) or those with higher tourism potential (Savoie vineyards). They also share a pressing issue related to the establishment and succession of vineyards, as well as attractiveness to winemakers and employees. Finally, they suffer from a lack of resources, both human and financial, to reverse these trends.

Nonetheless, new assets are emerging, which are also collectively shared. First, they can leverage their AOP, an official quality label, to emphasize a perception of qualitative distinctiveness. In addition, the trend towards discovering authentic wines, those that are lesser-known, rare, and potentially sought after, could work to their advantage. Finally, they can similarly highlight the unique assets of their individual territories in terms of landscapes, heritage and environment, which are all tied to their viticultural terroir. Thus, these vineyards face common challenges related to positioning, adapting to the market, promoting their wines and enhancing their collective visibility.

10.3.2. *Study background and objectives*

In the spring of 2016, the Vignobles Forez Roannais association issued a call for proposals, which was awarded in the summer to a collective formed within the Institut Agro de Montpellier, led by Foued Cheriet (full Professor) and Hervé Hannin (research engineer), in collaboration with a start-up hosted there, Via Terra (Clément Arnal).

The specifications outlined in the agreement that followed in the summer of 2016 were as follows:

> High-altitude vineyards, Côtes d'Auvergne, Côtes du Forez, Côte Roannaise, and Saint-Pourçain share a common history rooted in polyculture: small family estates used to either produce wine independently or deliver their grapes to the cooperative cellar. They also share, among other grape varieties, gamay, and a specific type of terroir characteristic of the Massif Central. Today, these vineyards are recognized through their respective appellations of origin, and a new generation of winemakers is eager to introduce their wines to the local residents and to raise awareness beyond the borders of Auvergne-Rhône-Alpes. Recognizing their common strengths and challenges, they have decided to take matters into their own hands and unite. Over the past three years, they have conducted joint press initiatives (hosting journalists, organizing the third edition of the Ici commence la Loire exhibition, and more). To enhance their visibility in the wine industry, they aim to define a list of shared and unifying values with the goal of establishing collective communication. To create this shared identity, they plan to conduct a study on the collective strategy for these four vineyards.

Beyond evident issues of common positioning and visibility, many other challenges are associated with this collective co-construction. First, there are challenges with respect to the collective approach: can we achieve more and better results by combining everyone's efforts? We will see that the belief in the feasibility of the project was forged throughout the year 2016–2017. However, it is accompanied by a set of success conditions that must be met because, as with any collective endeavor, the risks of deviation or stagnation should not be underestimated. It has helped clarify two questions: (1) What can we do based on the collective ambitions that have emerged? (2) How can we do it, by what path and with what priorities?

10.3.2.1. *A socio-economic co-construction*

We understand that the mathematics of a collective can become synergistic (4 + 1 = 5!) as long as we follow the following progression: getting to know ourselves better in order to confidently present ourselves to others and then gaining a deeper understanding of how our qualities and weaknesses are perceived. Most importantly, this process should be integrated into the project's dynamics: "how can my uniqueness be beneficial and drive the project?" and "how can my uniqueness, which may be seen as an obstacle to the project, be minimized or even balanced by the qualities of the other vineyards?".

10.3.2.2. *A strategic growth and value approach*

The purpose of this collective effort is undeniably strategic. It involves all the winegrowers for the long term. In addition, it is part of the growth strategy (e.g. aiming for greater visibility, improved competitiveness and better resource allocation) and is also tied to the value strategy. In this context, the early choice made by the winegrowers is "differentiation". Consequently, the question arises of how to make sure that this differentiation is recognized in the market, and the decision was to address it by seeking a relevant and distinctive common positioning that respects the specific attributes of each of the four vineyards.

10.3.2.3. *Marketing questioning*

The pursuit of a collectively developed common positioning, in the marketing sense of the term, appears to be a logical and essential prerequisite for presenting a united and coherent image in the "markets". There is no question of common achievements or even communication without a shared and thoroughly discussed platform. Thus, the question of building the positioning was presented as the core of the response to the call for tenders and a prerequisite for any operation.

Educational efforts were required to help the winegrowers accept and understand the methods used throughout the study, which were largely based on a marketing approach. This included market analysis (using survey methods and evaluating indicators such as awareness and the image of the vineyards), the development of a strategic positioning and the creation of fundamental tools for a communication strategy.

10.3.2.4. *An issue of visibility and expanding into markets, including international ones*

The transformation of French wine markets is well-documented. The continuous decline in consumption within our "traditional" markets necessitates export initiatives and the exploration of new markets in competitive environments. Consequently, the ability to generate internal resources and leverage interprofessional or local support is vital for entering international markets. The question of establishing strong local roots through geographical attributes and institutional connections is pivotal. Therefore, as is the case with most French vineyards, the central issue is how to structure an entity that is firmly rooted in its territory and has achieved a critical size. This entity should be well equipped with resources and capable of generating interest among customers and consumers.

10.3.2.5. *The territorialization question*

This raises the question of winemakers‘ ability to promote the project throughout the various bodies and channels in this extended territory. Transforming territorial organizations into project ambassadors becomes a paramount task. The challenge is more significant because, unlike other more renowned vineyards in France, the organizations and representatives of the cities and nearby territories rarely promote their own wines. The question of aligning the project with a recognized, visible and consumer-friendly territory arises. Likewise, the issue of establishing a connection, to be defined, with a large wine organization that would provide specific attributes and services without compromising the unique qualities of the four vineyards coming together arises.

10.3.2.6. *The leadership question*

As is the case in any collective project, identifying and building effective and inspiring leadership is necessary. The unique characteristics of the sponsor group's composition required adjustments in the steering groups to ensure the support and representation of winemakers. This observation further emphasized the choice to prioritize a well-structured organization with legitimate representatives.

These questions "being questioned" are part of the complexity of the approach.

Faced with the challenges they encountered, the four vineyards recognized the importance of a collective positioning, especially since they had already initiated some collaborations and were beginning to understand both the advantages and limitations. They gradually started to come together, such as in the merger of the

Forez and Roannais vineyards within the AVFR (Association des Vignobles du Forez-Roannais) in 2012, joint representation of the Saint-Pourçain and Côtes d'Auvergne vineyards at the International Agricultural exhibition, joint press initiatives and the organization of the "Ici commence la Loire" event, a national gathering of the four vineyards aimed at the press and wine professionals.

These collaborative efforts led to the formation of a "working group", composed of representatives from each vineyard, to collectively explore further opportunities. This working group commissioned the study, conducted between 2016 and 2017, in the form of a "research-accompaniment" type call for proposals. The selected proposal aimed to assist the representatives of the four vineyards in jointly developing a unifying and shared common strategy, focusing on three consecutive main objectives:

– the definition of common value(s) shared by all four vineyards, the basis of a common positioning on the wine market and a communication strategy intended to make it known;

– the appropriation of the joint project by winegrowers of each vineyard, to attract their support and guarantee the success of the collective approach undertaken;

– the definition of collective organizational modalities for carrying out and steering the project, via a unique structure, recognizable by the wine and territorial authorities and the prescribers of the world of wine.

The undertaken approach was ambitious: to facilitate the creation of a legitimate collective with a common position, capable of pursuing a distinctive strategic approach to generate value and contribute to the sustainability of vineyards through a new market positioning for wine. It is worth noting that within the project's dynamics, the nascent winemaker collective had to contend with "centrifugal" forces involved in the desired consolidation (such as the AVFR, the Puy-de-Dôme Wine Federation and independent winegrowers from Saint-Pourçain). In addition, there were "centripetal" forces at play: Agamy, a cooperative that includes Côtes du Forez, focusing on gamay valorization and alignment with Rhodanian vineyards; the Saint-Pourçain cooperative, somewhat in the background, emphasizing its own reputation in the Parisian market and the "natural" winegrowers of Côtes d'Auvergne, who were determined to remain independent.

10.3.3. *Methodological approach: data collection and analysis*

The issue raised was immediately recognized as multifaceted and complex, involving economics, sociology, marketing, agronomy and more. The goal was not simply to add up the attributes of each vineyard, but rather to establish a rigorous working methodology: (1) comprehensive enough to encompass all dimensions of the issue, (2) capable of fostering genuine collective ownership by the sponsors and (3) adaptable to the availability of as many participants as possible.

Throughout the study, it was necessary to balance the need to establish solid foundations for this project with the sense of urgency, at times strongly expressed by certain managers, to quickly produce visible results. Knowledge of sustainable projects in France and abroad has also inspired the ambition to build a new entity "over time". Thus, we had to strike a balance between these short-term and long-term perspectives in our approaches.

The research-accompaniment approach relied on extensive surveys to gather primary and secondary data, periodic visits to the collectives during the initial "introspective" phases and workshops for reflection and creation. Over the two years of the study, "stops" were conducted to present and discuss the preliminary results, co-create the next steps and refine intermediate objectives. A steering committee, which included winemakers, was established to oversee the process and validate the deliverables.

The goal was not to provide a ready-made, one-size-fits-all solution. Instead, the aim was to generate a strong consensus so that the representatives of the four vineyards could support and advocate for it among other winegrowers and local stakeholders. The success of the study and its anticipated impact largely depended on the wholehearted support of all stakeholders, ongoing ownership and strong facilitation skills to develop dissemination tools, create new resources and maintain a continuous connection between the project initiators and other stakeholders. The study followed a four-phase progression (see Figure 10.2): documentary analysis, surveys, data collection, group facilitation and report and recommendation writing.

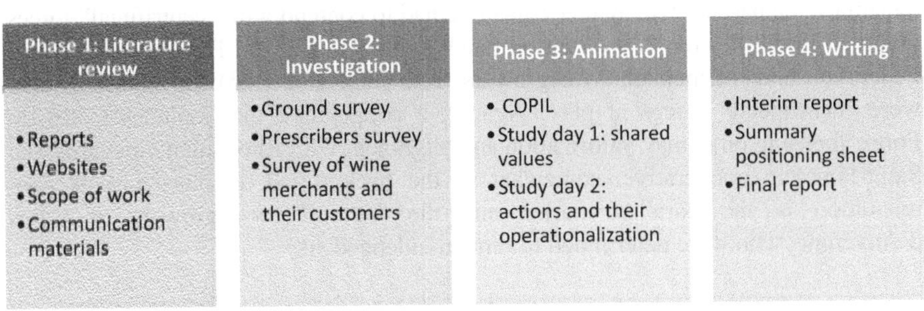

Figure 10.2. *Summary of the methodological approach. For a color version of this figure, see www.iste.co.uk/cheriet/wine2.zip*

A variety of diverse and complementary methodological tools were used due to their suitability in providing the required data and their capacity to inform the decisions of the representatives from the four vineyards. In this context, five surveys involving different types of respondents (see Table 10.3) enabled a comprehensive understanding of the issue. The summary results of these surveys are presented in Appendix 1.

Surveys	Sample and method of administration	Objectives
Paris visits	– Five restaurants/wine shops – SIAL Paris Stands visit	– Questionnaire test – Maintenance/image of the four vineyards and reputation
Producers survey	39 face-to-face interviews	– Mutual knowledge of vineyards – Appreciation of the collective project
Prescribers survey	– 22 "resource" people, very diverse profiles, wine or restaurant specialists – Self-administered questionnaire and additional phone-based interviews/follow-ups	– "External" image of the four vineyards – Relevance of the collective project – Assessments on the proposals of winemakers – Actions to be carried out
Wine merchants survey	– 33 wine merchants – Face-to-face interviews, eight cities in France	– Reputation of the four vineyards – Price and reference statements – Collective project appraisals and actions
Customer survey	– 35 on-site clients – Face-to-face interviews	– Reputation of the four – Relevance of the collective project
Consumer survey	– Online questionnaire, about 30 questions – 150 people	– Knowledge of the four vineyards – Image and reputation – Associated assessments

Table 10.3. *Summary of survey steps*

In addition, throughout the process, we initiated a co-construction and support process involving the stakeholders of these vineyards through numerous exchanges. This was part of a forward-looking reflection on their collective organization and strategic positioning in the wine market. In addition to the surveys conducted, this research was marked by the active presence of researchers for nearly two years, including six days of framing sessions, participatory workshops and intermediate and final feedback sessions.

10.4. Results and discussion: from four small vineyards looking for collective positioning to Loire Volcanique wines

10.4.1. *Dynamics of co-construction of collective positioning*

The co-construction of a shared collective positioning by the four vineyards relied on various tools used in the study.

Surveys conducted among influencers, wine retailers and consumers helped us understand how these external parties perceived the different vineyards (including their image, identity, associated values, etc.) and the idea of a collective project. These perceptions informed the thinking of the engaged winegrowers. The survey conducted among the winemakers and the first workshop were focused on mutual understanding: how each vineyard was perceived by its own winemakers and how winemakers from different vineyards perceived each other.

One of the objectives was to get to know ourselves and others better:[2]

– An initial workshop thus allowed us to identify common points, the foundations of a collective strategy, as well as stumbling blocks to work on to avoid any future animosity within the collective: for example, the opposition between Rhône-Alpes and Auvergne vineyards or between cooperative wineries and independent winemakers.

– The second workshop made it possible to work on the collective project, with the objectives to (1) specify the common elements to be retained and put forward to build a collective strategy, (2) determine how to recognize and appropriate this collective positioning within the various vineyards, (3) anticipate the actions to be undertaken to make it operational and (4) specify the governance to be adopted to manage it.

– Other small group workshops focused on communication actions, media and targets to focus on (see Figure 10.3).

---

2 The workshops established and conducted during these sessions benefited greatly from the active collaboration of Patrice Lallemand, an engineer from the Institut Agro Montpellier (IHEV), to whom the authors would like to extend their special thanks.

a)

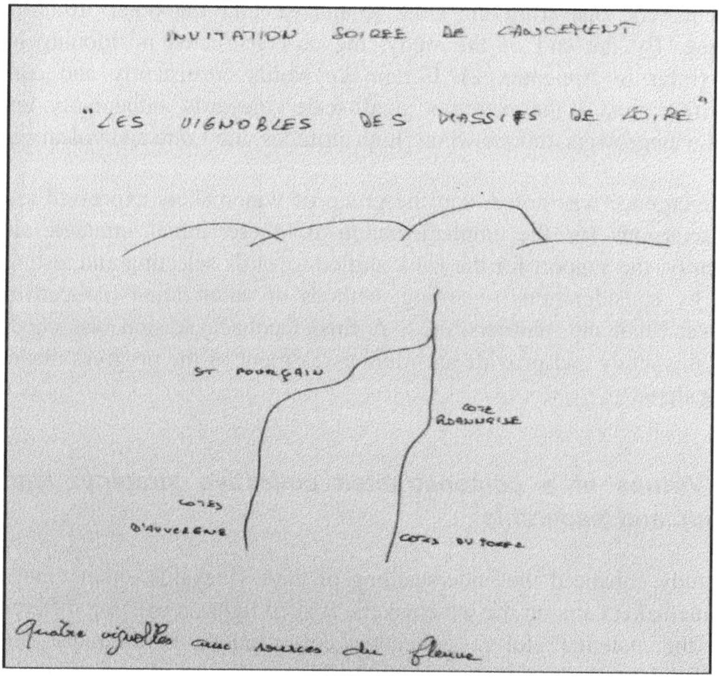

b)

Figure 10.3. *a) Workshop for the creation of event posters by winemakers. b) Example of a poster being created (photo credit © Cheriet F.). For a color version of this figure, see www.iste.co.uk/cheriet/wine2.zip*

This co-construction effort has unveiled several necessary conditions for the success of a future common positioning. Early on, it became apparent that there was a need to strengthen inter-vineyard cohesion, foster "closeness" among vineyards and winemakers, promote a culture of exchange and sharing to enhance mutual understanding and learn to speak with a unified voice. Moreover, the development of a collective positioning relied on a shared name and values that resonated with both the winemakers and the market. The collective strategy embraced the wholehearted involvement of winemakers closely tied to the project, ensuring their full ownership of it. The goal was not to dilute the distinct identities of the vineyards or erase their differences, but rather to celebrate each one's strengths in service of the collective. Finally, it was crucial to maintain equilibrium within the vineyards, ensuring that everyone had a role in the collective project and that no one felt marginalized.

Throughout all phases, we witnessed a palpable determination among the involved stakeholders and genuine capabilities to carry out the strategic exercise to its completion. As a result, a true leadership emerged within the collective, led by two winemakers, one from the Côte Roannaise and the other from the Côtes d'Auvergne. By the end of the study, the co-constructed positioning had been clarified (refer to Appendix 2). Certain keywords consistently and collectively emerged from most of the exercises: small-scale vineyards, authenticity, united and dedicated winegrowers, unique wines, high altitudes, the Loire and volcanoes.

A convergence was noted, and the group of winemakers expressed a desire to begin discussions on the implementation of these initial strategic decisions. Subsequently, the support for the work shifted towards selecting and testing names, followed by considerations regarding methods of valorization (collective brands, governance, financing, partners, etc.). A final feedback session was conducted to conclude this study and provide an initial assessment of the progress made and the upcoming steps.

10.4.2. *Virtues of a co-constructed collective strategy: legitimacy, ownership and leadership*

The study enhanced the understanding of their vineyards, other vineyards and fellow winemakers among the winemakers. It shed light on existing differences and revealed the potential for a collective construction. Furthermore, the study uncovered the strategic and operational challenges. It also identified shared values across the four vineyards, which could serve as a foundation for a meaningful collective positioning. These common values gained legitimacy through the representation of the engaged winemakers and the contributions of the various approaches used in the study.

Analyses of both internal and external perceptions, coupled with discussions involving vineyard representatives, have contributed to the foundation of this positioning. It is built upon shared elements of identity (the Loire, volcanoes and the Massif Central), a primary grape variety to be highlighted (gamay), a strong connection to local gastronomy and "human" values upheld by winegrowers (simplicity, authenticity, friendliness). These values are associated with their vineyards, which are characterized as small, authentic, human scale and known for producing original wines.

Embarking on a collective endeavor already showcased the vineyard representatives' clear awareness that a common strategy for all four vineyards would be quickly needed to secure their future. The various phases of this study illustrate the progress made in this strategic deliberation surrounding the challenges of such a collective endeavor. "Progress" is, without a doubt, the term that best characterizes the efforts exerted by the vineyard representatives in getting to know themselves, defining their characteristics, identifying their strengths and weaknesses, determining the key aspects of their project and anticipating or addressing the main obstacles. This was genuinely a co-construction in the entrepreneurial sense of the word, complete with moments of hesitation, back-and-forth and introspection.

In the end, such an approach has undeniably strengthened the ownership of this collective strategy, as its potential stakeholders were actively involved in its conception and eventual realization. Following the study, the process of structuring this collective effort continued, resulting in the establishment of Loire Volcanique on November 14, 2019, as a non-profit association: 43 winegrowers from the four vineyards united around the slogan "Between Loire and volcanoes, a preserved vineyard."

The claimed positioning is that of "authentic small vineyards": according to the press kit, Loire Volcanique represents "a unique history, an exceptional climate, and a central France geography, a collective of dynamic, supportive, and committed winemakers, biodiversity as heritage, gamay as a common cornerstone, and [...] diversity as strength". The marketing rollout began in 2020, with support from the Soubiran agency (see Appendix 3 for two examples of promotional material).

Communication actions were conducted by the collective to enhance its visibility and promote the joint project. This included collaborating with a service provider to create a graphic charter, a logo (see Figure 10.4) and a communication plan. The collective also launched a showcase website (see: http://loire), engaged in social media communication (e.g. the Instagram account @loirevolcanique), produced and shared podcasts on Spotify, carried out press and influencer campaigns, and organized events, such as annual tastings for professionals in Paris (2022 will mark the third edition).

Figure 10.4. *Logos of the four vineyards and that of the Loire Volcanique. For a color version of this figure, see www.iste.co.uk/cheriet/wine2.zip*

10.5. Conclusion

The results presented indicate that the process of constructing the collective positioning of the four vineyards was foundational and essential, albeit complex and marked by certain challenges. Nonetheless, the process was successfully carried out by the stakeholders, and the collective is now self-reliant and capable of implementing communication actions to showcase its distinctive characteristics.

Even though it was carried out as a collective entrepreneurial project (Émin and Schieb-Bienfait 2013), limitations related to divisions within and between vineyards, the choice of the management system (Schieb-Bienfait et al. 2020) and questions regarding leadership and project management (Yan and Sorenson 2003) likely hindered its implementation.

From a theoretical standpoint, the analysis of collective dynamics highlights the challenges of an exclusive conceptual framework (Cook and Plunkett 2006). Assessing it through the lenses of collective entrepreneurship and project management in specific territories aligns with the complex, evolving and geographically grounded nature of the analytical constructs. This alignment becomes even more pertinent as it involves dimensions of entrepreneurial orientation, leadership and governance of the collective project. The construction of this positioning was approached as an entrepreneurial project in its own right.

Methodologically, the progressive approach implemented over an extended period facilitated a comprehensive understanding of the situation. The various surveys involving winemakers, prescribers, customers and others, along with interactions during workshops and feedback sessions, contributed to the evolution of inquiries and in-depth exploration of specific challenges. We contend that this research-accompaniment approach (Bréchet et al. 2014) is an integral component of the co-construction process, acknowledging the potential limitations stemming from a potential loss of objectivity among the accompanying researchers.

Although this study is based on a single case, and we cannot assert the representativeness of the results or the replication of the methodological approach (Yin 2017), it is important to highlight the relevance of this research-accompaniment approach. It enables the analysis of collective entrepreneurial projects through a comprehensive and context-specific examination of the issues, thus avoiding the pitfalls associated with the diversity of actors and their objectives, as well as the complexities of the dynamics involved. Other inquiries led the collective to reject certain choices regarding governance (e.g. vineyard merger), financing (sole reliance on local government subsidies) or partnerships (integration of actors from outside the territory). This study provided the necessary foundation for the marketing initiatives initiated in 2019: the establishment of an association and a "brand", *Loire Volcanique*, and the engagement of a communication agency to develop various media (website, social networks, press kit) and plan and host promotional events. The "gestation" period of 2017–2018 was truly the birth of the collective and helped clarify its positioning.

This research raises significant questions about the future evolution of this collective endeavor. First, it prompts consideration of the conditions necessary to sustain the momentum through the formalization of governance and leadership. Second, it addresses the development of additional operational initiatives and the reinforcement of the collective positioning. Finally, questions arise regarding the establishment of connections, be it with other stakeholders in the region, other industries, institutions or other collectives and associations in neighboring territories with compatible or potentially synergistic positions, such as those promoting gastronomy in Roanne.

The results obtained also offer potential directions for future research. First, they raise questions about the analysis of the connections and compatibilities between the collective and individual positions of each vineyard. Second, future research could explore the broader analysis of the outcomes of collective positioning in the wine industry: what are the collective performance and impacts on each vineyard (Renting et al. 2011)? Third, another avenue for research could involve the examination of externalities in terms of regional and territorial development, tourism and the reach of collective positioning on national and international scales (Dana et al. 2014).

Finally, these results would be strengthened through a comparative analysis of the dynamics involved in constructing other collective positions in the wine sector in different regions (Loubaresse and Pestre 2016) or collective positions of other sectors within the same territory (Loire-Auvergne). Outlined in this way, these research directions provide an understanding of the importance of analyzing collective entrepreneurial phenomena within the wine industry and the potential offered by this research theme in terms of researchers' support in developing relevant, useful and context-specific collective strategies.

10.6. Appendices

10.6.1. *Appendix 1. Summary of survey results*

10.6.1.1. *Producer survey (39 winemakers, October 2016)*

– Marked identities of vineyards, and attachment of winemakers to their terroir and its specificities.

– Despite differences, many common values were identified: authenticity, simplicity, small vineyards, granite and volcanic terroirs, grape varieties, history, sectors, etc.

– Relevance of a collective project (unavoidable) in view of the common problems encountered by vineyards (aging of winemakers, generational renewal and maintenance of cooperative tools, positioning within the wine market, image and visibility, etc.): the need to find common solutions together.

– Common problem of the low reputation and poor image of wines, especially internally: the need to be appreciated by the local population and to make the actors of the territories "ambassadors" of the vineyards.

– Problem of differentiation vis-à-vis other vineyards, especially Beaujolais.

– Willingness of winemakers to be on a very operational project: collective communication actions, pooling of resources, purchasing group, etc.

– Willingness to unite all winegrowers in all vineyards and to "leave no one behind".

10.6.1.2. *Prescribers survey (22 influential actors, November 2016 to January 2017)*

– Relevance of a collective project (unavoidable) in view of the common points and specificities of vineyards.

– Marked specificities of vineyards (positive/negative points).

– Many common values were identified: authenticity, simplicity, small vineyards, nature, Massif Central, terroir and grape variety.

– Different dynamics according to vineyards: generational renewal, qualitative progression and implications of leading winemakers.

– Common problem of low reputation and poor image of wines.

– Common actions were identified, ranging from communication actions, events, media dedicated to professionals, to the creation of a common structure, a brand, a quality label, etc.

– The "wine sourced from Loire" name is considered irrelevant from a marketing point of view and in relation to the historical and geographical characteristics of vineyards.

– Geographical anchorage still uncertain (rather Massif Central, Center).

10.6.1.3. *Wine merchants survey (33 wine merchants, France, face-to-face interviews, February 2017)*

– Weak overall presence, but to be qualified, and differentiated referencing of the wines of the four vineyards. Presence consistent with the reputation observed in other surveys. Only two wine merchants (Le Puy-en-Velay) present an offering of wines from the four vineyards.

– 24 out of 33 wine merchants have at least one reference and presence of at least one wine merchant per city with an offering of these wines. Commercial arguments: originality, light and aperitif wines, and small vineyards. Fairly positive customer feedback and fairly high loyalty.

– Insufficient commercial activities and promotional events for the four vineyards. Wine merchants are supportive of a project involving a combined offering of the four vineyards, while preserving their distinct identities and establishing a shared entity as the point of contact for wine promotion and tasting events.

10.6.1.4. *Wine merchant client survey (35 on-site clients, face-to-face interviews, February 2017)*

– 60% of respondents had already tasted wine from at least one of the appellations.

– The vast majority of those who have never tasted these wines are curious to discover them.

– Reputation in line with what has already been noted (Saint-Pourçain, Auvergne, Roannais, Forez).

– Vocabulary associated with wines: light wines, terroir, rustic, nature, conviviality and authenticity (in order of recurrence).

– Clients believe that these wines have a bad image and an unstable quality.

– A majority of clients, including those who know at least one of the four vineyards, are in favor of a collective project: promotion of terroir/originality, joint communication, joint efforts to improve the quality of wines and joint presence at trade wine fairs.

– When it comes to a common name, individuals have proposed associations with the location: "Massif Central", "Center", "Loire", "Auvergne", and with the terroir: "volcano" and historical references: "vin des Gaulois" and "vins des rois".

10.6.1.5. *Electronic consumer survey (150 online consumers)*

– 57% of respondents know at least one of the vineyards.

– Perceived reputation is generally low and heterogeneous. This is in line with previous results and with decreasing scores: Saint-Pourçain, Auvergne, Roannais and Forez. Overall, 98% of respondents know about vineyards and wines: "bad" to "average" awareness.

– Related words: terroir, authenticity, mid-range, rustic, traditional and artisanal.

– A majority of those surveyed support a collective grouping/project. It is worth noting that 31% reject it.

– Common name (descending order of recurrence): “Center” (30%), “Auvergne” (24%), “Massif Central” (12%) and “Loire” (12%) and “volcano” (9%).

– Actions to be taken (descending order of recurrence): joint communication, wine tourism actions, joint promotions with wine merchants and restaurateurs, joint presence in trade shows.

– Three quarters of customers for a first discovery at wine merchants and restaurateurs.

10.6.2. *Appendix 2. Conclusion of the research-accompaniment work and suggested/desired positioning*

– Authentic small vineyards, dynamic winemakers, original wines, altitude, volcanoes and Loire.

– Customer promise: up-and-coming wines (waiting to be discovered), unique. Committed and dynamic winemakers, small vineyards taking charge of their destiny together (to be highlighted as a collective effort). Wines embodying values (small-scale, simplicity, authenticity, humility, etc.) that can add purpose to your consumption.

– Distinctive features: small but dynamic together. Wines waiting to be explored, unique, one-of-a-kind. Four vineyards, yet a wealth of wines to discover. “Altitude” wines, a defining characteristic of Massif Central wines.

– To highlight/emphasize in a communication plan: press relations and opinion leaders (at local and national levels), involvement of elected officials and local stakeholders. Storytelling: authenticity of winemakers (small-scale and committed), the compact size of the vineyards (a mark of quality and authenticity), engagement in a collective project.

– “Discovery” events (for professionals and customers): showcasing the diversity of wines, along with a connection to (local and gourmet) cuisine, landscapes and terroirs. Geographically covering the nation and major neighboring cities (Lyon, Saint-Étienne, Paris and Clermont-Ferrand). Potential wine tourism initiatives.

– Positioning in one sentence (name, collective brand): small vineyards, dynamic winemakers, authentic wines to discover in the mountains, volcanoes and massifs where the Loire originates, etc.

10.6.3. *Appendix 3. Communication materials. Loire Volcanique press kit (August 2022)*

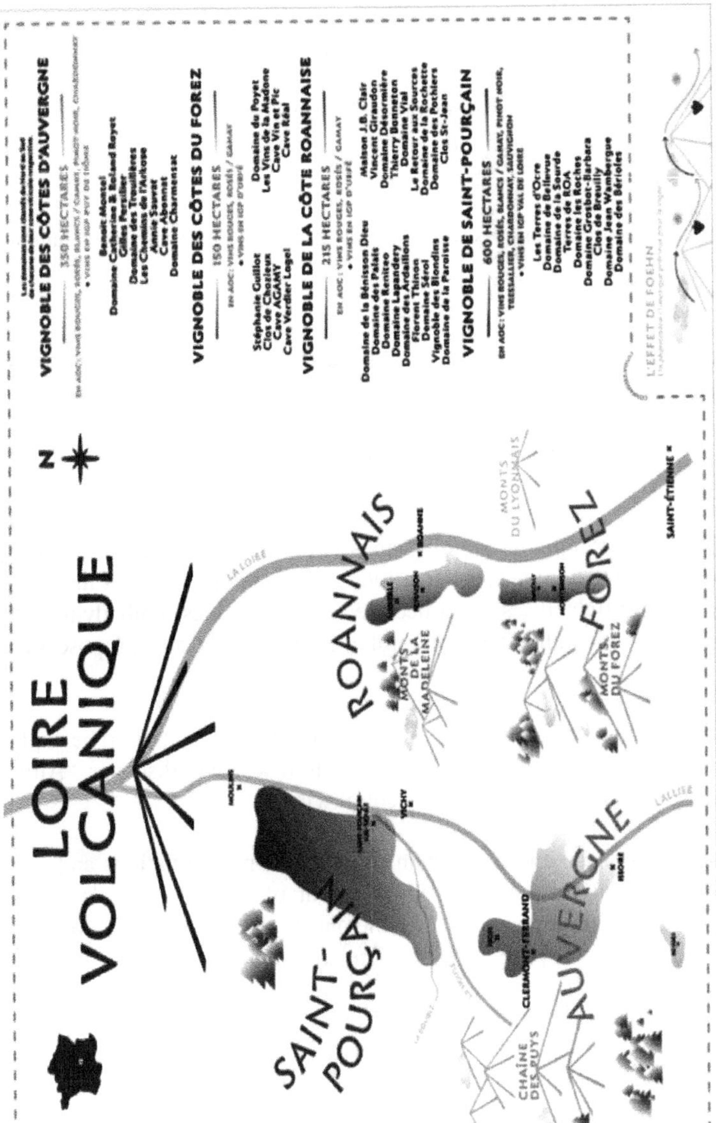

Figure 10.5. *Loire Volcanique communication visual 1 (August 2022). For a color version of this figure, see www.iste.co.uk/cheriet/wine2.zip*

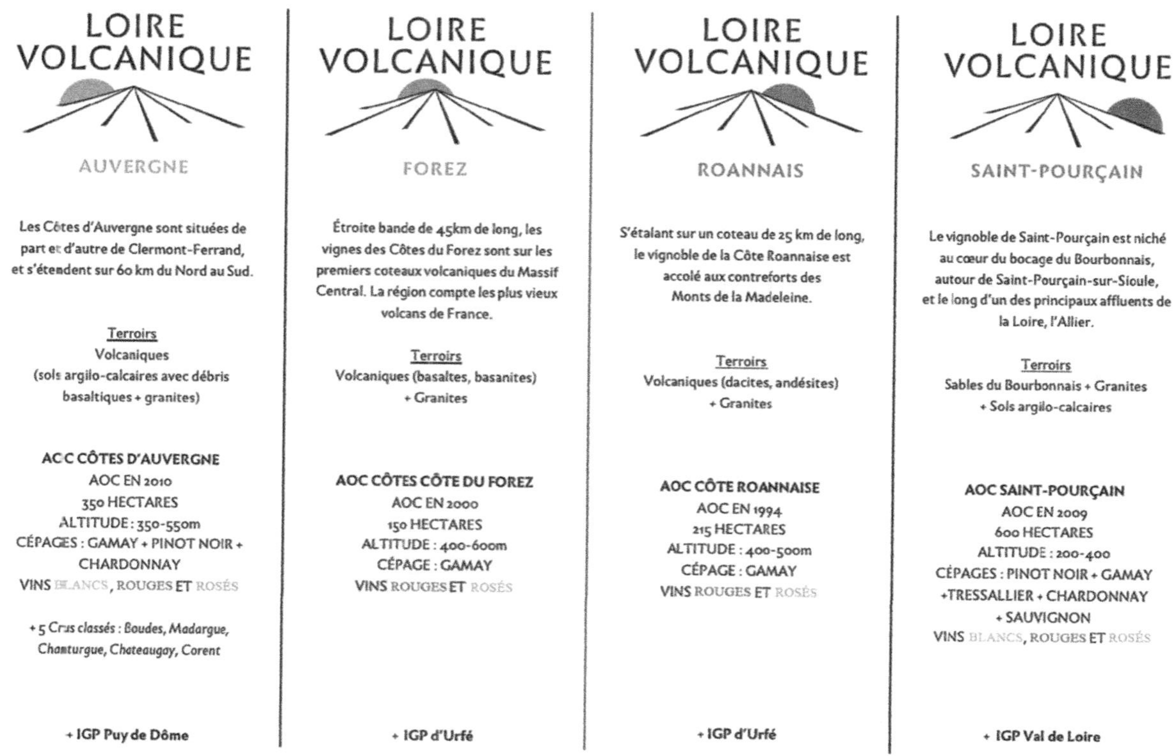

Figure 10.6. *Loire Volcanique communication visual 2 (August 2022). For a color version of this figure, see www.iste.co.uk/cheriet/wine2.zip*

10.7. References

Amblard, A., Berthome, K., Houdart, M., Lardon, S. (2020). L'action collective dans les territoires. Questions structurantes et fronts de recherche. *Géographie, Économie, Société-GES*. doi: 10.3166/ges.20.2017.0032, hal-02435040.

Blancanaux, R. (2019). Quand les instruments d'action publique (dés)inhibent de nouvelles stratégies collectives et économiques. L'exemple du vin de cépage en Languedoc-Roussillon. *Revue Politique Européenne*, 1(63), 36–60.

Bréchet, J.-P., Émin, S., Schieb-Bienfait, N. (2014). La recherche-accompagnement : une pratique légitime. *Finance Contrôle Stratégie*, 17(2). doi: 10.4000/fcs.1477.

Cook, M.L. and Plunkett, B. (2006). Collective entrepreneurship: An emerging phenomenon in producer-owned organizations. *Journal of Agricultural and Applied Economics*, 38(2), 421–428.

Dana, L.-P., Gurau, C., Lasch, F. (2014). Entrepreneurship, tourism and regional development: A tale of two villages. *Entrepreneurship & Regional Development*, 26(3–4), 357–374.

Émin, S. and Schieb-Bienfait, N. (2013). De la pertinence des approaches par le projet pour analyser les processus entrepreneuriaux collectifs : une étude de cas sur le territoire de la confédération paysanne. *Revue de l'Entrepreneuriat*, 1(12), 15–42.

Knickel, K., Zerger, C., Jahn, G., Renting, H. (2008). Limiting and enabling factors of collective farmers' marketing initiatives: Results of a comparative analysis of the situation and trends in 10 European countries. *Journal of Hunger & Environmental Nutrition*, 3(2–3), 247–269.

Loubaresse, E. and Pestre, F. (2016). Les leviers de performance d'une stratégie collective : une analyse des filières vitivinicoles en Val de Loire, en Champagne et en Languedoc-Roussillon. *Annales des Mines - Gérer et Comprendre*, 123, 46–59.

Martin, M. and Tanguy, C. (2018). Proximités et construction d'une stratégie collective : application au cas de deux filières agroalimentaires. *Revue Géographie Economie et Société-GES*, 4(20), 423–447.

Peyronnet, J.M. (2022). Vins de Loire : une identité culturelle à construire. *Cavistes & E-commerce*, 19, 30–32.

Renting, H., Shermer, M., Oostindle, H. (2011). Collective farmers' marketing initiatives in Europe: Diversity, contextuality and dynamics. *International Journal of Sociology of Agriculture and Food*, 18(1), 1–11.

Schieb-Bienfait, N., Dufeu, I., Bréchet, J.-P. (2020). Le dispositif de gestion, moteur d'une dynamique entrepreneuriale collective. Le projet Bio Loire Ocean. *Revue de l'entrepreneuriat*, 4(19), 57–79.

Yan, J. and Sorenson, R.L. (2003). Collective entrepreneurship in family firms: The influence of leader attitudes and behaviors. *New England Journal of Entrepreneurship*, 62(2), 37–51.

Yin, R.K. (2017). *Case Study Research and Applications: Design and Methods*. Sage Publications, Thousand Oaks, CA.

11

Research and Training, Factors of Competitiveness and Adaptation: The Montpellier Vine and Wine Cluster Case

11.1. Introduction

Viticulture has expanded globally by creating specialized production areas, whose names have become increasingly linked to the quality and identity of wines, even leading to the international emergence and proliferation of numerous geographical indications. The existence of these wine regions, varying in size, has long intrigued geographers, historians, economists and managers alike. What factors account for their locations, developments and often their achievements?

The initial explanations that emphasized natural factors (terroirs) or commercial factors (proximity to cities or commercial routes) (Dion 1959) were followed by analyses highlighting the organizational dimension of these territories (Ditter 2005). This included the role of associations or producer unions and, more broadly, "clusters" that integrated research and training organizations dedicated to viticulture and oenology (Porter 1998; Visser 2004; Boyer and Touzard 2016). In all the vineyards worldwide, the capacity of wine stakeholders in the same region to organize themselves plays a crucial role in approaching markets and influencing political contexts. It helps them defend and promote their wines, sustainably manage local resources that contribute to their quality and develop innovations that enable wine businesses to enhance their competitiveness and adapt to emerging challenges like climate change (Belletti et al. 2017; Hannin et al. 2020). In this chapter, we delve into the specific ways in which the relationships between wine companies and research and training organizations can be established within a region to address

Chapter written by Hervé HANNIN and Jean-Marc TOUZARD.

contemporary challenges in competitiveness and adaptation. To do this, we will first revisit the research that has tackled this issue in both established and emerging wine-producing regions (Part 1). We will then illustrate this using the case of the "wine and vine" research and training center in Montpellier and its evolving connections with the vineyards in Languedoc, now part of the Occitanie Region (Part 2).

11.2. Research, training and innovation: conditions for the competitiveness and adaptation of new and old vineyards

11.2.1. *Development of wine clusters in new wine-growing countries*

Since the late 1990s, a series of studies in economics and management science have highlighted the crucial role of research, training and innovation in the development of vineyards in "new world" countries. This includes regions like California (Porter and Bond 1999; Mueller and Sumner 2006; Hira and Swartz 2014), Argentina (McDermott 2007), Chile (Visser 2004; Giuliani and Bell 2005), South Africa (Lorentzen 2009), Brazil (Fensterseifer 2007) and Australia (Roberts and Ingram 2002; Aylward et al. 2006). In these countries, where European migrants established vineyards in the 18th and 19th centuries, wine production and exports have seen significant growth, particularly from the mid-1980s as economies became more internationalized (Coelho and Rastoin 2001). This expansion was made possible by lower production costs, foreign investments, supportive legislation and a simplified business approach to wine. However, all these studies also underscore the pivotal role of developing new knowledge and innovations, often facilitated by the geographical proximity between wine companies, professional organizations and universities or research centers.

Indeed, the significant presence of "new wine-producing countries" in global wine markets cannot be solely attributed to a process of productive specialization capitalizing on natural comparative advantages. It relies on endogenous factors associated with the territorial development of networks, expertise and institutions that closely integrate research and technical and higher education organizations. Universities such as Davis in California, Stellenbosch in South Africa, Adelaide in Australia, as well as Mendoza in Argentina and Santiago in Chile, are at the core of clusters (Porter 1998) where:

– engineers, technicians and salespeople are trained, who will implement the new viticulture;

– knowledge and advice is built and disseminated to take into account local production conditions;

– connections between these local networks and companies or research and training centers in other countries are established, particularly France;

– an entrepreneurial and innovative wine culture, often associated with regional business plans, is elaborated;

– projects are formalized in partnership with positive externalities, for example, on the development of workers' skills or companies' access to new technologies.

These studies on the territorial dimension of the competitiveness of new world vineyards align with the research that has been evolving in various sectors since the early 1990s, concerning industrial districts, localized production systems, clusters or regional innovation systems. They also emphasize the significance of university–research–business relationships for economic competitiveness, exemplified by the triple helix model (Etzkowitz and Leydesdorff 2000).

The pioneering vineyards in the new wine-producing countries broadly affirmed these analyses, which were often more focused on industrial and service activities within the "new knowledge economy" (Foray 2010). Competitiveness unfolds in the long run in territories that organize themselves, foster innovation and training, and engage in cooperation networks at both local and international levels.

11.2.2. *Analysis of wine clusters in France: a new French paradox?*

Paradoxically, research on the evolution and competitiveness of European vineyards has only recently begun to focus on territorial conditions of innovation and clusters. With a few exceptions, such as studies conducted in Italy in a comparative analysis with Chile (Morrison and Rabellotti 2009) and in France on AOC wine clusters (Calvet 2005) or localized agri-food systems (Touzard et al. 2008), research on innovation in European vineyards, especially in France, has traditionally emphasized two aspects. First, there has been research at the national sector level, analyzing its economic and institutional transformations (Hannin et al. 2010) or exploring potential scenarios for its evolution (Sebillotte et al. 2003). Second, there have been studies examining the adoption of innovations at the level of winegrowers and their professional networks within a vineyard (Compagnone 2004; Chiffoleau and Touzard 2014; Mascarenhas and Touzard 2018), without delving into regional research and training institutions.

However, in response to the research on wine clusters in the new world, studies began to emerge in France from 2014, particularly in Montpellier and Bordeaux, aimed at characterizing the regional innovation systems of vine and wine, and analyzing their impact on the competitiveness of vineyards (Hinnewinkel et al. 2015; Boyer and Touzard 2016). Their initial findings indicated varying relationships, to the extent of suggesting the existence of a French Paradox. At the beginning of the 21st century, the developments in French vineyards presented a stark contrast. On the one hand, there was dynamism in Champagne, Alsace and Burgundy, while on the other hand, there was a decline in the commercial position of Bordeaux wines and a continued undervaluation of wines from Languedoc-Roussillon, with Montpellier serving as its regional capital. Interestingly, these latter two regions had the highest number of publications, researchers and educators dedicated to vine and wine!

Does the Bordeaux and Languedoc-centric focus of French scientific research on vine and wine have minimal influence on the competitiveness of their regional vineyards? Nevertheless, a closer examination of research and educational mechanisms within French wine regions has helped to put this apparent paradox into perspective (Boyer and Touzard 2016):

– Calculating research intensity (employment or budget dedicated to research relative to the surface or production volume) partially addresses the previous observations. Research intensity is notably higher in the vineyards of Alsace, Burgundy and Champagne (with over three full-time equivalents of regional public research per 1,000 hectares in 2015), making them the most competitive vineyards in the 2010s. In contrast, Languedoc-Roussillon, covering nearly 250,000 hectares, exhibits low research intensity (around 1). The gap is somewhat less pronounced when considering research and development (R&D) full-time equivalents in relation to the volume produced, owing to the lower average yields in Languedoc-Roussillon (Boyer and Touzard 2016). Nonetheless, paradoxical situations persist, particularly in Bordeaux, which boasts high research intensity, and conversely, in the Loire Valley, a region experiencing economic growth, but displaying low apparent public research intensity.

– Clear effects of inertia or chronological discrepancies become evident between institutional changes (the reorganization of an interprofession, a new regional project, etc.), public or private R&D investments and the economic performance of a vineyard. This complicates the attribution of these investments to the competitiveness of a regional vineyard and raises questions about the capacity of its stakeholders to anticipate potential futures (Cheriet et al. 2020).

– On the other hand, a more detailed analysis of the R&D investments made by the interprofessions in each wine-growing region shows a positive impact on the economic dynamics of the vineyards. When an interprofession actively participates in R&D with dedicated budgets and personnel (similar to the Comité Champagne), it enhances the competitiveness of its vineyard by fostering experimentation and the generation of specialized knowledge. It also facilitates knowledge dissemination and the structuring of a "demand for research and innovation", which regional or non-regional research centers can address.

The research conducted on European vineyards, complementing earlier efforts in new wine-producing countries, highlights that a vineyard's competitiveness depends on several factors. These include the intensity of research and development (R&D) in its region, its capacity to anticipate and plan its actions, and, most importantly, the presence of effective regional governance guided by an interprofessional organization. This organization should be capable of defining research needs and establishing meaningful collaborations and knowledge networks.

11.2.3. *Towards innovation ecosystems to adapt to environmental and climate change challenges*

The research conducted on wine clusters in both new and traditional wine-producing countries encourages us to specify the organizational and relational conditions that go beyond strengthening their positions in the markets. These conditions are crucial for ensuring their adaptation to increasingly pressing challenges, such as preserving their natural environment or addressing climate change. It is no longer solely about promoting the adoption of technical, commercial or product innovations by fostering cooperative relationships between companies, professional organizations and research and training centers within each cluster. It involves a more comprehensive and strategic revision (Ollat et al. 2016; Hannin et al. 2020) and is based on at least four key dimensions:

– taking into account the evolutions of ecosystems, soils and climate, which can no longer be regarded as unchanging factors influencing wine quality, as traditionally associated with terroirs;

– opening up the governance of vineyards to other actors in the territory, involved in the management of local resources and adaptation to climate change;

– identifying and sharing experimentation, developing innovations and changes in more and more interacting areas of action (plant material, agronomic and oenological practices, spatial reorganization of vineyards and management of their ecosystems, risk management, certification and communication, etc.);

– combining these different levers of action in adaptive strategies at the local and regional levels, anticipating different paths and possible bifurcations, etc.

These strategic dimensions for vineyard adaptation align with recent research directions on territorial innovation dynamics, replacing the concept of clusters centered on a territory's productive activities with that of local innovation ecosystems (Torre and Zimmermann 2015; Scaringella and Radziwon 2018; Boyer et al. 2021). This concept seeks to provide a more comprehensive understanding of the evolving conditions for innovation, emphasizing the significance of both:

– multiple collaborations with other sectors and civil society;

– taking into account the dynamic and metabolic relationships with the local environment and its "natural ecosystems".

Several recent studies have sought to clarify the evolution of relationships between wine companies and training and research centers within local innovation ecosystems: in California (Levy and Lubell 2018), Italy (Chaminade and Randelli 2020) and Brazil (Marques et al. 2021).

In France, a comparative analysis of these relationships was also conducted among the wine regions of Bordeaux, Champagne and Languedoc. This analysis started with an examination of the factors that could explain changes in the perception and practices of winegrowers to adapt to climate change (Boyer and Touzard 2021). In these three vineyards, the commitment to adaptation strategies of wine companies is linked to three factors:

– the background and training of the managers of these companies;

– previous adoption of environmental approaches, including certification in organic viticulture;

– the presence of frequent and strategically important relationships (measured through an egocentric network analysis) with professional organizations (such as chambers of agriculture and interprofessions), environmental associations, or research centers and universities. These relationships notably involve hosting interns, which is one of the most significant explanatory variables.

This research validates the fact that the integration of new environmental issues, especially climate change adaptation, is significantly influenced by relationships established within the local or even regional innovation ecosystem of a vineyard. The key question is how these relationships are formed and how they can be promoted through strategic alignment among actors in regional wine governance: professional organizations, research and training institutions, businesses, government bodies and local authorities. The history of the "vine and wine center" in Montpellier serves as an illustrative example of the challenges and prerequisites for such development.

11.3. The Montpellier research and training center and the transformation of Languedoc viticulture

11.3.1. *Legacy: a corporate organization defending table wine, supported by the Montpellier School of Agriculture*

Teaching and research activities related to viticulture and winemaking began to flourish in Montpellier in 1872, with the establishment of its School of Agriculture. Montpellier found itself at the heart of a region that had become specialized in table wine production, thanks to the development of railways and in response to the growing national demand driven by urbanization and industrialization. The early decades of the School of Agriculture were first focused on researching the phylloxera epidemic, which had affected vineyards in the Languedoc and throughout France (Legros and Argelès 1986). Professors Ravaz and Foex, from the Viticulture department, collaborated with their peers at the Jardin des Plantes and the university (notably Planchon) to identify the aphid, understand the extent of its damage, trace the origins of the infected plants back to America and ultimately resolve the issue through widespread grafting onto rootstocks, some of which were of American origin. This successful endeavor quickly elevated the reputation of "wine science in Montpellier" beyond national borders. Certainly, this science was already a collaborative effort, involving many colleagues from various countries and regions, including individuals like Westwood in Oxford, Millardet in Bordeaux and Sahut in Languedoc, who were knowledgeable winegrowers, nurserymen and agronomists leading scholarly societies. These collaborations between wine industry notables and scientists already demonstrated the early formation of a wine cluster, driven by the urgency to address the phylloxera crisis. The publication of the magazine *Le Progrès agricole et viticole*, which started in 1884, initiated by regional winegrower families and guided by professors from the school's Viticulture department, serves as a notable example of this collaboration.

The post-phylloxera reconstruction of the Languedoc vineyard followed a model sometimes characterized as industrial (Augé-Laribé 1907). This approach benefited from the education and the new viticultural and oenological techniques experimented with and promoted by the School of Agriculture. However, it resulted in a crisis of overproduction, which reached its peak during the Languedoc events of 1907. This crisis prompted the government to introduce sector-specific regulations, distinguishing between, on the one hand, the production of table wines without yield limits, but subject to macroeconomic regulatory measures (such as distillation and uprooting), and, on the other hand, a terroir-based viticulture with a system of protected designations of origin (Bartoli and Boulet 1989). Languedoc soon adopted a political approach to managing the production and markets of table wine, overseen by the Confédération Générale des Vignerons du Midi (CGVM). This approach was gradually supported by a network of mutualist and cooperative institutions, including the federations of cooperative wineries in Gard, Hérault and Aude. Within this "cluster" of table wine production in the Midi region, close connections with the Montpellier School of Agriculture persisted, as evidenced by the significant "scientific and sometimes 'political'" involvement of three of its professors:

– Jules Milhau, an economics professor and the author of pioneering econometric studies on wine markets, advocated for the cooperative model. His commitment extended to chairing the Caisse de Réassurances Agricoles du Midi and later the Mutualité Sociale Agricole in 1948. He played a role in the establishment of the Compagnie Nationale du Bas-Rhône Languedoc (an irrigation company) in 1954 and supported the early efforts to restructure the regional vineyard. One notable contribution was his involvement in the creation of the Saint-Chinian appellation.

– Jean Branas, who served as a viticulture professor from 1930 to 1975, continued the research initiated by his predecessors on downy mildew. In 1949, he established the Vassal ampelographic conservatory in Marseillan-plage, using the School of Agriculture's vine collection as a foundation. He conducted numerous expert analyses and authored technical publications that would significantly impact the local wine industry stakeholders.

– Roger Lacombe, a machinery professor, played a significant role in the introduction of the first tractors in the Languedoc vineyard. In 1948, he established the *Journées de démonstration de motoviticulture* (Days of motoviticulture demonstration) in Montpellier, which later evolved into SITEVI in 1976. SITEVI is now an internationally recognized trade fair.

Following World War II, the establishment of new research, training and development institutions further solidified Montpellier's position in the modernization of regional viticulture.

– Upon its establishment in 1945, the Institut National de la Recherche Agronomique (INRA) set up on the campus of the Montpellier School of Agriculture, becoming the fifth center in France. Research efforts were firstly directed towards enhancing viticulture productivity, aligning with the School's educational programs and the needs of both winegrowers and businesses along the regional supply chain. Furthermore, the acquisition of the Pech Rouge estate near Narbonne in 1956 enabled INRA to establish an experimental unit in the western part of the regional vineyard.

– The Faculty of Pharmacy in Montpellier expanded its oenology programs in parallel with the School of Agriculture, which transformed into the École Nationale Supérieure Agronomique de Montpellier (ENSAM) in 1949. The recognition of the oenologist profession in 1955 led to the establishment of the National Diploma of Oenology (DNO), jointly awarded by both institutions. ENSAM grants the DNO to agricultural engineers specializing in viticulture and oenology, who undergo multiple internships in regional cellars and across French vineyards.

– Starting in the 1960s, both national and subsequently European agricultural policies aimed to enhance the modernization of agriculture through a cooperative approach involving the state, trade unions and agricultural organizations, with the assistance of agricultural chambers. In Languedoc, these chambers were responsible for providing viticultural advice, while oenological advice was offered by private laboratories or the Institut Coopératif du Vin (ICV), established in 1949. ENSAM and INRA supported these organizations by providing training and conducting technical experiments and study sessions.

Languedoc-Roussillon then emerged as the world's leading wine region, reaching its peak in 1970 with nearly 440,000 hectares and over 30 million hectoliters of wine sold, 80% of which was bulk wine for the domestic market. Besides 4,000 private wineries, 550 cooperative wineries accounted for 95% of the region's 120,000 winegrowers, who enjoyed incomes higher than the national average for agriculture. In this environment, the Languedoc wine cluster was highly competitive, driven by a robust corporatist movement led by the CGVM. Research and training organizations in Montpellier were informally associated with this movement.

11.3.2. *Research and training during the reconversion and fragmentation of the Languedoc vineyard (1976–2000)*

Starting in the 1970s, the landscape underwent rapid changes. National consumption of table wine saw a sustained decline, while the European common market expanded, with Spain joining in 1982, making it more competitive. Crises within the table wine market became frequent, requiring significant political interventions, such as distillations and the uprooting of vineyards, both temporarily and permanently. Confronted with these crises, continuing a defensive strategy in table wine production seemed like a dead end. This realization followed a series of violent and unsuccessful protests, culminating in the tragic events of Montredon in 1976, which resulted in casualties among both the demonstrators and law enforcement. In response, wine industry leaders from the south began to establish connections with educators and researchers in Montpellier. The goal was to gain a deeper understanding of the crisis and explore alternative paths for "restructuring the southern vineyards" (Touzard and Laporte 1998):

– New technical approaches are being explored to support the vineyard's restructuring based on two models. On the hillsides, regional grape varieties (Grenache, Mourvedre, Syrah, etc.) are carefully chosen to establish AOC. In the plains, emblematic grape varieties from other regions (Merlot, Cabernet Sauvignon,

Chardonnay and more) are being planted to develop “local wines” that are promoted as “varietal wines”. This shift in the vineyard’s focus benefits from guidance from educators like Denis Boubals. It necessitates the acquisition of new knowledge about these grape varieties and their cultivation methods and encourages the development of new grape varieties, as suggested by Paul Truel (Marselan, Caladoc, etc.). These endeavors also involve experiments related to mechanized harvesting and fermentation control.

– In addition, contacts are being established with social science educators and researchers who are investigating the evolution of wine farms, cooperative wineries and shifts in wine consumption trends in France, a pivotal element in the crisis. A study initiated in 1980 by INRA Montpellier and ONIVIT (National Interprofessional Office of Table Wines) concerning wine consumption behaviors (Boulet and Huguet 1980) is playing a crucial role in shedding light on the strategic adaptations of Languedoc wine organizations. This study has evolved into a quinquennial survey, which has been ongoing for four decades and is currently overseen by FranceAgriMer (Aigrain et al. 1991).

In the 1990s, the restructuring of regional viticulture gained momentum. Technical systems began to focus strongly on yield management, companies underwent transformations, particularly cooperative cellars through mergers, and the wine supply became more segmented in response to market developments. The diversification of wines and strategies gave rise to new wine organizations, often leading to conflicts: AOC unions versus Pays wine unions, federations of cooperative wineries versus federations of private wineries run by independent winemakers, tensions between competing territories, etc. To coordinate the stakeholders in vine and wine within each wine region, interprofessions were established in areas with geographical indications starting in 1994. However, unlike other French regions that organize around a single interprofession (e.g. Champagne, Bordeaux, Burgundy), Languedoc-Roussillon ended up with as many as seven interprofessions: the Interprofessional Council of Languedoc Wines (CIVL) (representing AOCs in Hérault and Aude), the Interprofessional Council of Roussillon Wines (CIVR) (representing AOCs in Roussillon), InterRhône (representing AOCs in East Gard), Inter OC (representing wines from Pays d’Oc), the Association Nationale Interprofessionnelle des Vins de Table (ANIVIT) (representing table wines and other regional wines), the Comité Interprofessionnel des Vins Doux Naturels (CNVDN) (representing natural sweet wines) and the Association Interprofessionnelle des Vins Biologiques du Languedoc-Roussillon (AIVB-LR), founded in 1991 to promote the interests of organic viticulture. Although these interprofessions were more dispersed and less coordinated compared to those in other regions, their connections with the Montpellier campus persisted, albeit becoming less tight and less efficient.

The 1990s marked an era of internationalization in the wine markets. The Languedoc vineyard expanded its exports, facing competition from the "new world wines". Beyond the market competition, these newcomers challenged established quality standards. Strengthened by the Marrakech agreements and the establishment of the WTO, they even posed a threat to the traditional definition of wine and its production methods, encompassing both oenological and viticultural aspects (Hannin et al. 2006). In response to these changes, the wine industry leaders in Languedoc became increasingly interested in the OIV and used its global congresses to gain insights into the emerging world of wine. They visited the *Ko-operatiewe Wijnbouwers Vereniging* (KWV) in South Africa, sometimes finding themselves astounded by stockpiles of wood chips ready to be immersed in the vats. They also heard KWV's leaders proclaim their transformation from a mere cooperative into a "full-fledged company". In Australia, they met professionals who were planting "own-rooted vines", explaining that their *business plan* was built on a 20-year horizon, beyond which other, more profitable opportunities may develop.

Recognizing the strategic importance of developing education and research in wine innovation and marketing, the Languedoc profession, particularly young cooperators, turned to ENSAM, INRA and occasionally the "training center" of the International Organization of Vine and Wine (OIV), which had been offering such programs since 1986. Simultaneously, leaders within the Montpellier hub and enlightened professors started to explore this area, following the example of Daniel Boulet (INRA), who established the Centre Interdisciplinaire d'Études sur le Vin (CIEV) in 1988. In addition, Philippe Lacombe and Guy Albagnac lent their support to the short-lived Higher Institute of Vine and Wine in Montpellier (ISVVM).

As a result of repeated discussions with the OIV's "training" division (Michel Bourqui), the leadership of ENSAM (Paul Raynaud) and that of the INRA center (Daniel Boulet) found themselves pondering several crucial questions: what does the future hold for oenology education if the range of accepted practices aligns with those of the agri-food industry? Should we initiate research in wine marketing, even though INRA lacks both laboratories and researchers in this field? How can ENSAM maintain its position as a leading institution in viticulture and oenology? How can we improve the coordination of research on wine conversion and proactively address the sector's needs, especially in the absence of a representative organization? In response, three decisions made in the late 1990s paved the way for new opportunities in research and education, with the aim of revitalizing the Languedoc-Roussillon vine and wine cluster:

– Leveraging opportunities for research co-financing provided by the regions through decentralization laws, INRA initiated a research program in 1995 in partnership with a focus on agriculture and regional development (Permanent Delegation for Agriculture, Development and Foresight, DADP). In

Languedoc-Roussillon, this program supported various projects, including studies on uprooting and replanting dynamics, the adaptation of specific grape varieties (Syrah), changes in viticultural and oenological practices, as well as the strategies of agricultural enterprises and wine cooperatives, all aimed at fostering innovation for new types of wines.

– The leadership of ENSAM invited Hervé Hannin from the OIV to join them in establishing the Institute for Higher Education in Vine and Wine (IHEV) in Montpellier. Their goal was to develop new, more diversified courses on vine and wine that were open to agronomy and, most importantly, marketing. These courses were built on new partnerships with local and global sectors and universities.

– ENSAM initiated a national consultation with major national public wine organizations. This effort convinced INRA to launch a prospective study to explore the potential futures of the wine sector. The study aimed to highlight the fragmentation of organizations and their divisions, which posed challenges for developing a common strategy and anticipation in terms of research. Patrick Aigrain (Office National Interprofessionnel des Vins, Onivins) and Hervé Hannin (ENSAM) led this national prospective study under the guidance of Michel Sebillotte. Over the next 20 years, they conducted about 15 additional studies to assist sector organizations in shaping their strategies.

11.3.3. *A vine and wine cluster in Languedoc relaunched by educational and research institutes (2000–2014)*

Starting from the 2000s, viticulture in Languedoc-Roussillon stabilized after a period of significant vine removal, reducing the vineyard to less than 300,000 hectares (–33% compared to 1970) and production to 18 million hectoliters (–40%). Replanting efforts allowed for the growth of both AOC wines and, more significantly, country wines, especially the wines from Pays d'Oc that established themselves in the international varietal wine market. While regional viticulture organizations remained divided, initiatives from the state, the region and especially the research and training institutes in Montpellier began to multiply, aiming to structure a new wine cluster.

The creation of IHEV in 2004, a decision made by director Étienne Landais, allowed ENSAM to introduce eight training programs at the master's and professional degree levels over the course of a few years. These programs welcomed nearly 300 students, with one-third being international students. Notably, this expansion included new programs in economics and management, such as a master's and a professional license in wine marketing. The initiative for these programs came from the Montpellier agricultural high school (in collaboration with the University of Montpellier). In addition, IHEV introduced a program designed to prepare students for the international OIV diploma, the OIV MSc in wine management. The creation of IHEV also paved the way for multidisciplinary approaches to teaching, focusing on themes like "sustainable viticulture" and "wine and health" – this was significant as organic and sustainable viticulture practices were starting to gain traction. IHEV actively engaged in continuing education, played a role in organizing conferences at the SITEVI International Exhibition for vine-wine (Salon international des équipements et savoir-faire pour les productions vigne-vin, olive, fruits-légumes), olive, fruit–vegetable production and provided expert assessments, thereby strengthening and reestablishing connections between the campus, organizations connected to viticulture (IFV, chambers of agriculture, ICV, appellation unions, etc.) and the local winegrowers. ENSAM's economics and management faculty, collaborating with IHEV, also authored a series of educational books on wine markets from 2003 to 2007, which proved valuable to stakeholders in the wine sector, particularly in Languedoc (Bachus collection, Dunod).

The prospective study conducted by INRA between 2000 and 2003 on French viticulture also played a crucial role. It was subsequently adopted by IHEV and FranceAgriMer for a new exercise focused specifically on Languedoc-Roussillon's viticulture. This initiative followed a request from the Regional Directorate of the Ministry of Agriculture (Hannin et al. 2011). The study resulted in four scenarios for the year 2025, which were instrumental in shaping a regional wine strategy. One key driver for implementing this strategy was the Languedoc-Roussillon winegrowing basin council, an advisory body established by the state in 2006. In 2011, this council organized a two-day seminar based on the prospective study, with one of the main conclusions being the need to consolidate existing interprofessions or at least align their strategies. While efforts were made to achieve this goal, such as the inclusion of departmental PGIs into the CIVL in 2012 and collaborative actions between the CIVL and Inter OC, unification was not attained. Simultaneously, various other foresight and strategy development exercises were carried out for different stakeholder groups within the regional wine sector. These included efforts to organize suppliers through the Vinseo association, which was established in 2007 with support from the region. Similarly, actors in organic viticulture were restructured under the Sud Vin Bio association. These initiatives also extended to the Chamber of Agriculture of Hérault as it developed its departmental agricultural plan (PADH).

Furthermore, the new research projects initiated through INRA's DADP program have enabled the renewal of collaborations with viticultural organizations, especially CIVL and the Fédération Régionale des Coopératives. With the support of the region, these collaborations have introduced new thematic areas, such as exploring the relationship between viticulture and the environment (with a focus on reducing the use of chemical inputs). In addition, research efforts have expanded internationally, involving European projects and contributing to the work of organizations like OIV and Giesco, a research consortium whose chairmanship has been undertaken by viticulture professors at the institution. Initially led by Alain Carbonneau, it is now headed by Laurent Torregrosa, who has been the director of IHEV since 2021.

This dynamic, often personified in the past by "prominent figures" who were highly visible within the school (Denis Boubals), INRA (Claude Flanzy) or the university (Suzanne Brun), had to adapt, starting in 1998, to the creation of Joint Research Units (UMR) that brought together researchers and teacher-researchers from various organizations (INRA, ENSAM, University, Cirad, etc.) to collaborate on common themes. While this initiative certainly strengthened the scientific aspect of the viticultural and winemaking theme, it spread across about a dozen UMRs primarily defined by their respective disciplines (agronomy, economics-management, genetics, soil sciences, etc.). Only one UMR explicitly focused on vines and wine, the UMR Sciences for Oenology. Throughout the 2000s, this academic fragmentation of research had to grapple with the fragmentation of wine organizations, which at times raised questions about whether vine and wine research was still a prominent focus in Montpellier!

However, efforts have been made to structure and provide better exposure to vine and wine research for professionals. This includes the annual organization of the Journées Scientifiques Vigne et Vins de Montpellier, the continuation of the *Le progrès agricole et viticole* journal led by Alain Carbonneau until 2010 and the development of shared institutional documents (Blondin et al. 2015). Meanwhile, in Bordeaux, collaborative initiatives between INRA, the region, the university and the École Nationale d'Ingénieurs des Travaux Agricoles (ENITA) resulted in the establishment of Bordeaux's Institute of Vine and Wine Science (ISVV) with an impressive facility. This Bordeaux-based progress gained further strength in 2014 when the university received the IDEX label along with new public funding. Anne-Lucie Wack, the new director of Montpellier SupAgro (established in 2007 through the transformation of ENSAM), and Laurent Bruckler, president of the INRA center in Montpellier, took steps to enhance the coordination and visibility of Montpellier's "vine and wine" research and training hub (Bruckler and Wack 2014). Their proposal received a positive response within the context of the reorganization of the University of Montpellier and the merger of Montpellier Universities 1 and 2 in 2015.

11.3.4. *Towards building an innovation ecosystem to support the adaptation of a plural viticulture*

The year 2014 marks a significant turning point in the history of the Languedoc-Roussillon vine and wine cluster. The region merged with Midi-Pyrénées to form the Occitanie Region, with Toulouse as its capital. The new regional administration is pursuing a supportive policy for both agriculture and viticulture, as well as for research, concentrated in Toulouse and Montpellier. Occitanie's viticulture has stabilized, covering approximately 270,000 hectares, with an average production of 14 million hectoliters. In 2019, sales amounted to nearly 1.7 billion euros, with over 50% coming from IGP (Indication Géographique Protégée) Pays d'Oc grape variety wines, 30% from AOP (Appellation d'origine protégée) and 20% certified in organic viticulture (DRAAF[1] Occitanie 2022). Irrigation is on the rise in the vineyards (20% of its area in 2020), thanks in part to the "Aqua Domitia" project. The value of Occitanie wines, while still relatively low, is gradually increasing, benefiting from the diversity of its range and likely improved coordination among its organizations. However, the wine industry is significantly affected by climate change, leading to reduced yields and impacting the income of winegrowers due to heatwaves (2019), late frosts (2021) and more pronounced drought (2022).

In this context, the "vine and wine" research and training activities, which continue to be centered in Montpellier, are undergoing restructuring as part of the Montpellier Université d'Excellence (MUSE) university consolidation, which obtained the I-Site label in 2017, along with new funding. MUSE subsequently established the Key initiative Vine and Wine Sciences (KIM VW), bringing together all teachers and researchers working on vine and wine in the Montpellier area. This new academic momentum, INRA's commitments (reaffirming vine and wine as a priority at the Montpellier center) and the influence of Montpellier SupAgro, under the leadership of Bruno Blondin, director of the IHEV since 2017, have enabled the gathering of nearly 250 scientists from 10 UMRs, three experimental units (Pech Rouge, Vassal, Chapitre) and engineers from the Institut Français de la Vigne et du Vin (IFV). The new funding from MUSE is then directed towards:

– the signing of agreements and the organization of exchanges with internationally recognized universities or research centers in the field of vine and wine (such as Davis and Cornell in the United States, Stellenbosch in South Africa, INTA[2] in Argentina, Adelaide in Australia, etc.);

– the formation of partnerships with regional wine organizations, especially interprofessional bodies, to encourage co-funding of studies and research. For instance, the interprofession for Pays d'Oc wines (InterOc) is committed to ongoing research funding to gain a better understanding of the factors that affect the competitiveness of its wines and the strategies that can enhance their adaptability to climate change.

---

1 Regional Directorate for Food, Agriculture and Forestry.

2 Instituto Nacional de Tecnologia Agropecuaria.

KIM VW (Key initiative Vine and Wine Sciences) outlines four primary areas of focus: climate change (mitigation and adaptation), addressing societal expectations (particularly related to the environment), aligning quality with market demands and embracing the digital transformation. Research efforts, spanning various disciplines and interdisciplinary collaboration, are intensifying in the realms of ecological concerns, irrigation, precision viticulture and the development of disease-resistant grapevines as eco-friendly alternatives to pesticides. Genetics plays a significant role in these efforts and is closely connected to ampelography, which continues to be a prominent expertise in Montpellier.

The initiatives undertaken within KIM VW encompass training efforts, such as the Vine & Wine summer school held in 2022 for doctoral students. They also encourage collaborative communication regarding the courses available, involving partnerships with the IHEV, whether they are standalone or joint programs with the university's DNO. The research activities align with multiple projects addressing these same challenges on both national and international scales. These projects leverage participatory approaches to facilitate the development of novel viticulture strategies:

– The drive for climate change adaptation is notably spearheaded by the INRA's Laccave project (2012–2021), jointly led by Montpellier (Jean-Marc Touzard) and Bordeaux (Nathalie Ollat) (Ollat et al. 2016). This project exemplifies the researchers' ability to mobilize across disciplines and highlights the significance of collaboration among different wine-growing regions. Through the development of a long-term vision for 2050, complemented by participatory forums and extensive media exposure, the project forges stronger connections between researchers and wine industry organizations, especially within the Occitanie Region.

– Similarly, research units in Montpellier collaborated with other French centers and professionals in 2020 to initiate the VITAE project, co-led by Bordeaux (François Delmotte), Montpellier (Hervé Hannin) and Dijon (Marielle Adrian). This project focuses on exploring the potential for "pesticide-free production". Right from the beginning, the project was integrated into a national and European context (Dijon Conference, June 2022, European outlook).

– The advancement of digital technology in viticulture is a third major theme that has been developed through the DIGITAG convergence institute and the OCCITANUM future investment project. This effort is further supported by the establishment of the "Mas Numérique" led by Montpellier SupAgro. Researchers and educators within KIM VW are actively engaged in collaborations with emerging sector actors, including start-ups. In these collaborations, they contribute fresh insights and, at times, a critical perspective.

– In addition, several new European projects are addressing these themes on a broader scale, specifically focusing on Mediterranean vineyards. These projects cover topics such as irrigation (PRIMA Talanoa project), the preservation of natural resources (Oenomed) and the sharing of innovative solutions for adapting to climate change (Vineas.net platform).

Furthermore, this momentum played a role in the OIV's decision in 2020 to suggest to Montpellier the revival of the international ampelography course. This proposal was based on the world-class expertise of Jean-Michel Boursiquot and a highly effective collaboration between Montpellier SupAgro, INRA and IFV.

After developing training programs in the field of vine and wine (2004–2015) and facilitating the consolidation of research efforts (2016–2021), Montpellier SupAgro is enhancing its collaborations with businesses in the wine sector. Building on historic relationships established over decades and nurtured by the IHEV, particularly through its *alumni* networks, a "Vigne et Vin" corporate chair was established in 2021 with the support of KIM VW. This initiative brings together around 10 major companies in the sector. The new chair supports research activities by providing funding for internships and theses, as well as awarding a doctoral prize. It also organizes events such as seminars, workshops and think tanks for students, researchers and industry professionals in the regional viticulture sector. The goal is to facilitate knowledge exchange, stimulate inquiries, structure questions and inspire new research, training and transfer projects. The establishment of this chair, combined with the educational activities supported by KIM VW, also resonates on a national scale, thanks to the creation of the Agro Institute, which, starting in 2021, unites French agronomic institutions in Rennes-Angers, Dijon and Montpellier. This offers a valuable opportunity for fostering collaborations with the Burgundy cluster and beyond.

These capabilities to unite, work collaboratively across disciplines and engage in partnerships led KIM VW to respond to the "Défis d'Occ" call for projects in 2021, initiated by the Occitanie Region. The "Vinid'Occ" project was accepted, built on the notion that the region can capitalize on the significance of its educational and research activities, as well as the diversity within its vineyards and business strategies. This collective effort aims to establish the region as the innovation hub for future viticulture and oenology, fostering closer ties with select research units in Toulouse.

Since 2014, all these developments have built the components of a "regional wine innovation ecosystem" (Boyer et al. 2021), driven by KIM VW and the research support policies of the Occitanie Region and the state:

– The themes of research and training have expanded, in particular towards the challenges of the environment, adaptation to climate change and digital innovations.

– Collaborations between researchers and wine companies bring together stakeholders at different levels (regional, interregional and international), in particular with the best universities recognized on these themes.

– From 2026, a building primarily funded by the Occitanie Region will accommodate the IHEV, featuring a research unit (SPO[3]) that will extend its influence across all UMRs and incorporate shared spaces designed to host wine-related businesses.

– Within the region, approaches are more participatory and open to new actors, linked to water management or the environment, as well as to local authorities, such as the metropolis of Montpellier, which is co-financing this building and developing a new wine hub.

Nonetheless, the establishment of this innovation ecosystem, driven by research and higher education, still faces a significant challenge: reinforcing collaborations and, ultimately, enhancing coordination among the professional wine organizations themselves.

11.4. Conclusion

Our literature review of clusters in both emerging and traditional wine-producing countries has uncovered several conditions that elucidate the role of research and training in enhancing the competitiveness and adaptability of vineyards. This even suggests a transition toward local innovation ecosystems that are open to other sectors and society, while also addressing the sustainability issues of their regions. An examination of the history of the Montpellier vine and wine cluster and its surrounding area further supplements this analysis, bringing forth new inquiries.

The establishment of a wine cluster, encompassing research and training activities, does not necessarily stem from recent awareness of innovation challenges, as earlier studies on "new world" vineyards may imply (Porter 1998). A cluster can be rooted in a long history marked by several cycles of construction, stability and restructuring. In the case of Montpellier, the formation of a cluster traces back to the late 19th century, as a response to the phylloxera crisis. Subsequently, it adopted a more closed approach, focusing on a corporatist management of table wine, driven by a unitary organization. However, it later fragmented during a wine market crisis and subsequently underwent a series of restructurings, primarily facilitated by the region and institutions of higher education and research.

3 Sciences for oenology.

The factors that influenced the evolution of the Languedoc cluster are also evident in other situations examined in the literature, although they may not have the same historical depth. These factors include the significant influence of the "inherited wine structure" (such as the type of vineyard management, grape varieties, wines, markets and location), which explains the unity and subsequent fragmentation of the governance of the Languedoc vineyards. Other factors involve the role of crises in mobilizing stakeholders, the impact of external economic (consumption, competition), technological and, notably, political factors, encompassing regional and national sectoral and scientific policies (such as the effect of MUSE on Montpellier). Last but not least, the initiatives and collective decisions made within the cluster, driven by wine industry leaders, scientists or political figures, play a crucial role in shaping the organization and governance of the cluster.

With regard to this organizational dimension of the cluster, the Languedoc example confirms that it is fundamental to the cluster's competitiveness, and is composed of factors mentioned in the literature, in particular on the "new wine-growing countries":

– the training connections of notable individuals, leaders, engineers and innovators who transform and adapt the regional vineyard;

– constructing a shared vision for the future of the regional vineyard was a unified effort during the "golden era" of table wine. Subsequently, it became necessary to rebuild this vision, primarily through strategic foresight exercises, to facilitate the transition;

– developing and connecting collaborations at different local/regional and international scales for better knowledge sharing;

– the role of partnership projects, generating multiple externalities related to the networks formed, the media coverage of results or associated learning;

– the importance over the last period of work on the environment and more open to society, illustrating the notion of "local innovation ecosystems".

But, beyond these common conditions, other points can be mentioned from the Languedoc case, enriching international work on wine clusters:

– The development of interdisciplinarity is crucial for addressing the increasingly complex challenges; in the face of academic and disciplinary norms, new projects, requests for expertise from local actors and the establishment of initiatives like KIM VW or the business chair help foster this interdisciplinarity.

– Beyond international connections, the development of interregional relations, including those between wine clusters (even within the same country, as in the case of Bordeaux and Montpellier), is advantageous for effectively managing "coopetition" relationships and enhancing collaboration within international organizations like the OIV.

– The advantages offered by a cluster are never guaranteed and require an ongoing effort of adjustment and anticipation, which was notably absent in Languedoc-Roussillon during the 1980s, but has been rekindled in recent years through foresight initiatives led by professional and scientific actors.

– Finally, the Languedoc example brings up the issue of how to capitalize on a diversity of strategies, resources, products and knowledge. This diversity carries inherent risks of fragmentation and conflicting interests and organizations, which can hinder unified governance. Conversely, it can offer opportunities for complementarity, openness to innovation and adaptability, highlighting potential sources of variety-based economies within a cluster that is based on diverse viticulture (Touzard and Laporte 1998). The Languedoc example illustrates the challenge of building these synergies, which, despite the incentives from scientific and political actors, have not yet been fully realized by wine organizations today.

The question then arises whether various forms of governance for a wine cluster (or a local innovation ecosystem) are feasible, going beyond the model reliant on professional management carried out by a single organization, akin to interprofessions in other French wine regions. A polycentric governance system, based on multiple territories and wine categories, could be trialed in Occitanie, facilitated by the coordination and substantial engagement of scientific and political stakeholders. Over the past 150 years, they have frequently succeeded in establishing adaptive synergies among their research themes, educational programs and projects, as well as their personal networks directed towards the diverse facets of this plural viticulture. This plural viticulture is now considered a potential common regional goal (Hannin et al. 2011).

11.5. References

Aigrain, P., Boulet, D., Lambert, J.L., Laporte, J.P. (1991). La consommation du vin en France : évolutions tendancielles et diversité des comportements. *Revue de l'Economie Méridionale*, 39(155–156), 21–52.

Augé-Laribé, M. (1907). *Le problème agraire du socialisme. La viticulture industrielle du Midi de la France.* Giard et Brière, Paris.

Aylward, D., Glynn, J., Gibson, B. (2006). SME innovation within the Australian wine industry: A cluster analysis. *Small Enterprise Research*, 14, 42–54. doi: 10.5172/ser.14.1.42.

Bartoli, P. and Boulet D. (1989). Dynamique et régulation de la sphère agroalimentaire : l'exemple de la sphère viticole. PhD Thesis, Université Montpellier I, Montpellier.

Belletti, G., Marescotti, A., Touzard, J.-M. (2017). Geographical indications, public goods, and sustainable development: The roles of actors' strategies and public policies. *World Development*, 98, 45–57.

Blondin, B., Hannin, H., Simonneau, T., Touzard, J.-M., This, P. (2015). Vigne et vin. Working Paper, Agropolis International, 21.

Boulet, D. and Huguet, J.-Y. (1981). La consommation du vin en France : enquête INRA-ONIVIT 1980. Report, Analyse des vins, des entreprises et des circuits/Ministère de l'agriculture/Office national interprofessionnel des vins de table/Institut national de la recherche agronomique, station d'économie et sociologie rurales.

Boyer, J. and Touzard, J.-M. (2016). Recherche, innovation et compétitivité des vignobles français : analyse par les institutions. *Systèmes alimentaires/Food Systems*, 1, 69–96 [Online]. Available at: https://hal.inrae.fr/hal-02629861.

Boyer, J. and Touzard, J.-M. (2021). To what extent do an innovation system and cleaner technological regime affect the decision-making process of climate change adaptation? Evidence from wine producers in three wine clusters in France. *Journal of Cleaner Production*, 315, 1–13.

Boyer, J., Ozor, J., Rondé, P. (2021). Local innovation ecosystem: Structure and impact on adaptive capacity of firms. *Industry and Innovation*, 28(5), 620–650. doi: 10.1080/13662716.2021.1891407.

Bruckler, L. and Wack, A-L. (2014). Lettre de mission pôle Vigne et Vin à G. Lépine et H. Hannin. Internal note.

Calvet, J. (2005). Les clusters vitivinicoles français à AOC : une analyse en termes de biens clubs. *Revue d'Économie Régionale & Urbaine*, 4, 481–506. doi: 10.3917/reru.054.0481.

Chaminade, C. and Randelli, F. (2020). The role of territorially embedded innovation ecosystems accelerating sustainability transformations: A case study of the transformation to organic wine production in Tuscany (Italy). *Sustainability,* 12(11), 4621.

Cheriet, F., Maurel, C., Amadieu P., Hannin, H. (eds) (2020). *Wine Management and Marketing: Opportunities for Companies and Challenges for the Industry*. ISTE Ltd, London, and John Wiley and Sons, New York.

Chiffoleau, Y. and Touzard, J.M. (2014). Understanding local agri-food systems through advice network analysis. *Agriculture and Human Values*, 31(1), 19–32.

Coelho, A. and Rastoin, J.-L. (2001). Globalisation du marché du vin et stratégies d'entreprise. *Économie Rurale*, 264–265, 16–34.

Compagnone, C. (2004). Agriculture raisonnée et dynamique de changement en viticulture bourguignonne : connaissance et relations sociales. *Recherches sociologiques et anthropologiques*, 3, 103–121 [Online]. Available at: https://hal.inrae.fr/hal-02678509.

Dion, R. (1959). *Histoire de la vigne et du vin en France ; des origines au XIXe siècle.* Flammarion, Paris.

Ditter, J. (2005). Clusters et terroirs : les systèmes productifs localisés dans la filière vitivinicole. *Reflets et perspectives de la vie économique*, XLIV, 35–51. doi: 10.3917/rpve.444.0035.

DRAAF Occitanie (2022). Données statistiques de la viticulture régionale [Online]. Available at: https://draaf.occitanie.agriculture.gouv.fr/vignes-et-vins-r576.html.

Etzkowitz, H. and Leydesdorff, L. (2000). The dynamics of innovation: From national systems and "Mode 2" to a triple helix of university-industry-government relations. *Research Policy*, 29(2), 109–123.

Fensterseifer, J.E. (2007). The emerging Brazilian wine industry: Challenges and prospects for the Serra Gaúcha wine cluster. *International Journal of Wine Business Research*, 19(3), 187–206.

Foray, D. (2018). *L'économie de la connaissance.* La Découverte, Paris. doi: 10.3917/dec.foray.2018.01.

Giuliani, E. and Arza, V. (2009). What drives the formation of 'valuable' university – industry linkages? *Research Policy*, 38(6), 906–921.

Giuliani, E. and Bell, M. (2005). The micro-determinants of meso-level learning and innovation: Evidence from a Chilean wine cluster. *Research Policy*, 34(1), 47–68.

Giuliani, E., Morrison, A., Pietrobelli, C., Rabellotti, R. (2010). Who are the researchers that are collaborating with industry? An analysis of the wine sectors in Chile, South Africa and Italy. *Research Policy*, 39(6), 748–761.

Hannin, H., Codron, J.M., Thoyer, S. (2006). The International Office of Vine and Wine (OIV) and the World Trade Organization (WTO): Standardization issues in the wine sector? In *Agricultural Standards: The Shape of the Global Food and Fiber System*, Bingen, J. and Busch, L. (eds). Springer, Berlin, 73–92.

Hannin, H., Couderc, J.P., D'Hauteville, F., Montaigne, E. (2010). *La vigne et le vin.* La Documentation française, Paris.

Hannin, H., Duquesnois, F., Aigrain, P., Ferrage, M., Escudier, J.-L. (2011). Perspectives d'évolution de la filière vitivinicole dans la région Languedoc-Roussillon à l'horizon 2025. Report, Ministère de l'Alimentation, de l'Agriculture, de la Pêche, de la Ruralité et de l'Aménagement du Territoire/FranceAgriMer.

Hannin, H., Touzard, J.-M., Aigrain, P., Bois, B., Brugière, F., Duchêne, E., Garcia de Cortazar-Atatauri, I., Gautier, J., Giraud Héraud, E., Hammond, R. et al. (2020). The French vineyard in the face of climate change: Developing an adaptation strategy based on prospective scenarios. In *Wine Management and Marketing: Opportunities for Companies and Challenges for the Industry*, Cheriet, F., Maurel, C., Amadieu P., Hannin, H. (eds). ISTE Ltd, London, and John Wiley and Sons, New York, 27–45.

Hinnewinkel, J.-C., Corade, N., Velasco-Graciet, H. (2015). *Essai géographique sur la crise du Bordeaux.* Maison des Sciences de l'Homme d'Aquitaine, Pessac.

Hira, A. and Swartz, T. (2014). What makes Napa Napa? The roots of success in the wine industry. *Wine Economics and Policy*, 3(1), 37–53.

Legros, J. and Argelès, J. (1986). *La Gaillarde à Montpellier*. ENSA, Montpellier.

Levy, M. and Lubell, M. (2018). Innovation, cooperation, and the structure of three regional sustainable agriculture networks in California. *Reg. Environ. Change*, 18, 1235–1246.

Lorentzen, J. (2009). Co-evolution in the western cape wine industry. Workshop: The Role of Innovation and Science in the Wine Industry, University of Novara.

Marques, K., Lermen, F., Gularte, A., de Magalhães, R., Danilevicz, Â., Echeveste, M. (2021). Inside of an innovation ecosystem: Evidence from the Brazilian wine sector. *Australian Journal of Grape and Wine Research*, 27, 66–80. doi: 10.1111/ajgw.12461.

Mascarenhas, G. and Touzard, J.-M. (2018). The social construction of quality in agri-food localized systems: The case of the Montpeyroux wine arrangement, France. *International Journal of Sociology of Agriculture and Food*, 24(2), 275–302 [Online]. Available at: https://hal.inrae.fr/hal-02622646.

McDermott, G. (2007). The politics of institutional renovation and economic upgrading: Recombining the vines that bind in Argentina. *Politics & Society*, 35(1), 103–143.

Morrison, A. and Rabellotti, R. (2009). Knowledge and information networks in an Italian wine cluster. *European Planning Studies*, 17(7), 983–1006. doi: 10.1080/09654310902949265.

Mueller, R. and Sumner, D. (2006). Grape and wine clusters. *Third International Wine Business Research Conference*, Montpellier.

Ollat, N., Touzard, J.-M., Van Leeuwen, C. (2016). Climate change impacts and adaptations: New challenges for the wine industry. *Journal of Wine Economics*, 11(1), 139–149.

Porter, M.E. (1998). Clusters and the new economics of competition. *Harvard Business Review*, 76(6), 77–90.

Porter, M.E. and Bond, G. (1999). The California wine cluster. Harvard Business School Case, 799–124.

Roberts, P. and Ingram, P. (2002). Vertical linkages, knowledge transfer and export performance: The Australian and New Zealand wine industries, 1987-1999. Unpublished manuscript, Emory University, Atlanta.

Scaringella, L. and Radziwon, A. (2018). Innovation, entrepreneurial, knowledge, and business ecosystems: Old wine in new bottles? *Technological Forecasting and Social Change*, 136, 59–87.

Sebillotte, M., Aigrain, P., Hannin, H., Sebillotte, C. (eds) (2003). *Prospective : vignes et vins : scénarios et défis pour la recherche et les acteurs*. INRA, Paris.

Torre, A. and Zimmermann, J. (2015). Des clusters aux écosystèmes industriels locaux. *Revue d'économie industrielle*, 152, 13–38. doi: 10.4000/rei.6204.

Touzard, J.-M. (2000). Coordination locale, innovation et régulation, l'exemple de la transition vin de masse – vin de qualité. *Revue d'Economie Régionale et Urbaine*, 3, 589–605.

Touzard, J.-M. and Laporte, J.-P. (1998). Deux décennies de transformation viticole en Languedoc-Roussillon : de la production de masse à une viticulture plurielle. *Pôle Sud*, 9, 26–47.

Touzard, J.-M., Chiffoleau, Y., Dreyfus, F. (2008). Analyser l'innovation dans un Système agroalimentaire localisé : construction interdisciplinaire en Languedoc. *Cahiers Agricultures*, 17(6), 526–531.

Visser, E. (2004). A Chilean wine cluster? Governance and upgrading in the phase of internationalization. *CEPAL*, 156.

12

Regional Wine Brands and Wine Brotherhoods: What Synergy to Promote Wine and Wine Tourism?

12.1. Introduction

Whether they are wine-related, Bacchic or vinic, the brotherhoods have undergone significant changes in recent years (Frochot 2003). Initially established for religious and spiritual purposes in France, wine brotherhoods have witnessed a transformation in both their numbers and roles, now extending to an international scale. They have shifted from being craft corporations or guilds primarily focused on safeguarding and regulating the skills and interests of their artisans[1] to becoming a public relations tool with the aim of promoting the wines of their region and enhancing awareness and comprehension of winemaking processes (Garrier 1999). In recent decades, wine brotherhoods have also started playing a role in the context of tourism development (Frochot 2003), as indicated by the increasing membership and popularity of their events (such as gala dinners and tastings of old vintages).

The Confrérie Saint-Étienne (CSE) is one of the oldest wine brotherhoods in Alsace. Its primary mission is to preserve and promote the Alsace vineyard. This institution fulfills multiple roles, serving as a historical and tourist monument, a guardian of Alsatian wine traditions and heritage, and a meeting point for wine enthusiasts of Alsace and local winegrowers. From a marketing perspective, the CSE can be viewed as a brand (Aaker 2012). It incorporates all its essential elements and key tools: a distinct identity, a rich history with a focus on longevity, a strong connection to the region, tangible heritage through its "wine library" and intangible

Chapter written by Coralie HALLER and Nada MAANINOU.

1 Larousse (1990). *Wines and Vineyards of France*. Edbury Press, London.

elements, such as songs and rituals and an engaging storytelling approach, along with diverse promotional activities, including Sigille[2] and various events.

There is another significant tool for promoting Alsatian vineyards: the Vins d'Alsace appellation. This is an AOC (AOP) created in 1962 and protected by the French government, recognized at the European level. From a marketing perspective, the Vins d'Alsace appellation also functions as a brand with specific guidelines, legal protection, production and labeling standards, distinct values and a unique identity, along with a visual charter (logo, colors) used for promoting wines from the region. This appellation is affixed to all wine bottles produced in Alsace that meet its specifications, allowing wines to be marketed under a unified regional identity.

In summary, both brands share common objectives, which are the valuation and promotion of the Alsatian vineyard. They target two main groups: Alsatian wine producers and their current and potential consumers. This leads to questions about potential collaboration and, more specifically, the synergy they can develop, particularly in terms of marketing practices and brand management. Establishing such synergy requires an analysis of how they are perceived by their respective target audiences (brand image). The aim is to mutually strengthen each other by leveraging shared positive perceptions and addressing any negative perceptions. As a result, we formulate three research objectives as follows:

Objective 1: investigate the brand image, reputation and role of the "Confrérie Vinique Saint-Étienne" among wine professionals (both members and non-members) and customers outside the wine sector.

Objective 2: explore the image of the "Vins d'Alsace" brand among wine producers in the region.

Objective 3: identify potential synergy effects between the two brands, based on the analysis of their respective images, for the promotion of wine and wine tourism in Alsace.

Box 12.1. *Research objectives*

To address these three objectives, we will first provide a literature review on wine brotherhoods and regional wine brands. Next, we will explain our research methodology, including an overview of the research context, followed by two case studies: one on the "Confrérie Vinique Saint-Étienne" (Objective 1) and another on

2 Every year, the CSE issues a quality label called "Sigille des Vins d'Alsace", rewarding the best Alsatian wines for their quality and distinctiveness (https://www.vinsalsace.com/fr/gouts-et-couleurs/sigilles/) [Accessed September 15, 2022].

the AOC Vins d'Alsace (Objective 2). Afterwards, we will present and discuss the findings from these case studies. Finally, we will discuss managerial recommendations for each of the two brands under study and the potential synergy between them (Objective 3). Our discussion will conclude with an integrative model on the synergy effects between the two brands.

12.2. Literature review

12.2.1. *Bacchic societies: origin and roles*

The creation of wine brotherhoods dates back to the Middle Ages in France, when the first guilds or professional corporations were established to protect wine-related professions, including winemakers, wine manufacturers, vineyard owners, taverns and sommeliers. During this period, religion, authority and corporatism were closely intertwined, and the spiritual and religious roles of wine brotherhoods were evident in various ways, such as appointing representatives to attend guild festivities and religious services, managing relief funds and associating the wine brotherhoods with a Bacchic brotherhood; this was derived from Bacchus, the god of wine in Roman and Greek mythology. In 1791, the brotherhoods were dissolved in France by a parliamentary vote. However, after the economic hardships that ensued World War II, they were revived to support and promote wine, sales and exports from their respective regions (Garrier 1999). They also aimed to preserve and pass on the culture of wine, including aspects such as wine tasting initiation and food and wine pairing. Wine brotherhoods are recognized as custodians of their local culture, which permeates all aspects of their existence, including their mission statements, values, rituals, ceremonies, rules, members, organizational structures, governance, events and attire.

Today, wine brotherhoods have been established worldwide with the goal of promoting wine and local culture, and supporting local economic development. In 1964, the Fédération Internationale des Confréries Bachiques (FICB)[3] was founded to network the numerous wine brotherhoods around the world and to support their activities and sociocultural values, including conviviality, the preservation of cultural and viticultural traditions, and the harmonious pairing of wine and food.

Despite their growth and significant role in promoting wine, wine brotherhoods have not received sufficient attention from academic researchers. Few studies have explored how these heritage-rich institutions can serve as marketing tools in collaboration with other actors in the wine and wine tourism sector, within a specific region. Our research is the first to (1) examine the image and role of wine

3 See: http://www.winebrotherhoods.org/en/ [Accessed September 15, 2022].

brotherhoods; (2) alongside a regional wine brand; (3) with a focus on promoting wine and the region as a wine tourism destination.

12.2.2. *The regional wine brand*

The region of origin of a brand is the region that consumers mentally associate with it, shaping their perception (Charters and Spielmann 2014). Just as the country of origin is important for some products, the region of origin is crucial for wine. Wine is generally closely connected to a specific place (region) and cannot be separated from it (Menival and Charters 2014). Research has shown that specific places, i.e. regions, add more value to wine brands than less specific locations (Thode and Maskulka 1998). This is partly because "regions are much more consistent in terms of human and natural factors than countries and can provide a more coherent, tangible, unique, and authentic image" (House and Bruwer 2003, p. 57). Wine is inherently tied to its region of origin, incorporating its geographical and cultural characteristics. The regional designation serves the same purpose as a brand, adding value to both the producer and the consumer (Ruzzier and de Chernatony 2013). From this research, we concluded that regional wine brands are closely associated with the region and provide added value to the wine.

In addition to these studies, literature on the concept of synergy suggests that the association between a territorial brand and agri-food products can lead to synergistic value creation effects (Lopes et al. 2018). In our case, the territory corresponds to the region, and the agri-food products refer to wine. Therefore, the synergy between the regional brand and the wines produced in this region can result in value creation.

The literature on place branding (Kavaratzis and Kalandides 2015) demonstrates that the region of origin can indeed be considered a brand in itself (Charters and Spielmann 2014). A *regional brand* is established when a region takes on the role of a brand, and in the case of wine and other related products associated with a place, becomes an umbrella brand[4] for the individual products and producers within it. A regional wine brand corresponds to the reputation of a wine-producing region in terms of its recognized and specific style of wine production (Easingwood et al. 2011, p. 21). For example, Barossa Valley, Australia's most well-known wine region and South Australia's most visited tourist destination, has built its success on European food and wine cultural traditions, and is known as an "Old New World" region[5]. Regional wine is a complex product due to the diversity of grape varieties

4 Brand used simultaneously for a set of heterogeneous products.

5 The South Australia Wine Story (2019, p. 13) [Online]. Available at: https://www.pir.sa.gov.au/__data/assets/pdf_file/0006/274065/South_Australian_Wine_Story.pdf [Accessed: September 15, 2022].

and terroirs within the same region, as well as the different perceptions consumers may have of the same wine. In this research, we focus on the brand image of regional wine and consider it to align with the classic definition of a brand image, meaning "the associations given to a wine brand by consumers" (Aaker 2012).

In summary, the literature discusses the concept of a regional brand, specifically in the case of wine, and highlights the benefits of synergy between the regional brand and agri-food products (in our case, wine). We propose to enhance this synergy by introducing a new player, the wine brotherhoods. These regional entities function as brands and work towards promoting the wines of their region and the region itself as a wine tourism destination. They have the potential to significantly contribute to value creation. From this perspective, it is crucial to explore and compare how these brotherhoods and the regional wine brand are perceived. To do this, we have selected the Confrérie Saint-Étienne (CSE) and the AOC Vins d'Alsace, supported by the Conseil Interprofessionnel des Vins d'Alsace[6] (CIVA), as our study subjects[7].

12.3. Methodology

12.3.1. *Context*

12.3.1.1. *The Confrérie Saint-Étienne (CSE): at the core of Alsace's wine heritage*

The CSE is regarded as one of the oldest in France and notably the most renowned, primarily because it owns a seat at the Château de Kientzheim, a splendid lordship built in 1564.

In terms of governance, the CSE is managed by a Grand Conseil (Grand Council) consisting of around 10 renowned professionals in Alsace wines, including winemakers and individuals dedicated to wine promotion through their respective roles. The council is presided over by a Grand Master, appointed for a one-year term, who introduces a specific theme for his tenure, notably realized through the organization of four annual solemn chapters. A general manager is responsible for executing the decisions of the Grand Master and Grand Council. The Youth Council, comprised of members of under 40, also provides support, particularly in activities aimed at promoting the CSE. Furthermore, admission to the CSE is granted following

---

6 The CIVA's role is to develop, promote and serve the Alsatian vineyard as a Vins d'Alsace regional brand.

7 The authors are members of the "Wine and Tourism" corporate chair at EM Strasbourg (https://www.em-strasbourg.com/fr/recherche/chaires/vin-et-tourisme). The CSE and the CIVA are partners of this chair.

wine-related tests upon the recommendation of a fellow member and approval from the Grand Council. To become a member, a person must progress through three ranks: (1) Apprentice Brother, followed by (2) Companion Brother after a one-year apprenticeship, and finally (3) Master Brother after a year as a companion. These ranks correspond to the individual's level of expertise in Alsace wines.

Figure 12.1. *The Saint-Étienne brotherhood: a) Kientzheim Castle; b) vault; c) logo; d) members of the Grand Council in their official attire; e) wine library (photo credit © Saint-Étienne brotherhood[8]). For a color version of this figure, see www.iste.co.uk/cheriet/wine2.zip*

8 See: https://confrerie-st-etienne.alsace/.

The CSE boasts a wine library, serving as a genuine repository of Alsatian vineyard history. Established in 1949, it houses over 65,000 bottles, with the oldest dating back to 1884. Among its prized possessions is the Mequillet[9] collection, featuring 600 bottles from the vintages of 1834 to 1837. Managed by the Young Adults Council of the Brotherhood, the wine library has been regularly replenished with "Sigillé" wines twice a year since 1957. Each distinguished winemaker contributes 12 bottles, each one carefully stored in microperforated plastic pouches. The CSE organizes prestigious annual auctions, with the proceeds contributing to the maintenance of the castle. The wine library thus serves as a tangible heritage of the brotherhood and a symbol of its strong reputation and renown. In addition, the CSE benefits from an intangible heritage characterized by its rituals (governance system, member appointments and ranks, solemn chapters) and the preservation and transmission of ancestral values during various events, such as workshops, ceremonies and auctions. These traditions are upheld both at the castle and internationally through approximately a dozen international delegations (including Belgium, Denmark, Germany, Great Britain, Ireland, Japan, Netherlands, Sweden, Switzerland and the United States). The solemn chapter event combines a ceremonial aspect with a gala dinner prepared by a renowned chef and exceptional Alsatian wines. Moreover, the CSE hosts "uncorking and recorking" workshops where bottles older than 30 years are opened to replace the "angels' share",[10] followed by the addition of sulfur to aid the wine's preservation before recorking. Special thematic or vintage-themed events also include the uncorking of bottles over 10 years old. These events underscore the importance of preserving Alsace's viticultural heritage and transmitting its values of sharing and conviviality with joy and respect for traditions.

However, the leaders of the CSE are aware of the need for rejuvenation and increased appeal, particularly among young winemakers and members, even though they express a strong attachment to ancestral traditions and operational methods. Case study 1 offers insights to address these issues by analyzing the perceptions of the CSE's target audience, thereby addressing the first objective of the research. In the following sections, we introduce the Alsace wine region and its AOC, Vins d'Alsace.

12.3.1.2. *Alsatian wine region and the AOC Vins d'Alsace*

Alsace produces 1 million hectoliters of wine annually, equivalent to 133 million bottles, with 90% being white wines. The wine is produced by around 3,812 producers and 750 wine-related businesses, primarily as part of cooperatives, as

9 Family of patrons in Alsace (https://www.alsace-histoire.org/netdba/mequillet/) [Accessed September 15, 2022].

10 Volume of wine that evaporates from the bottle over the years.

grape or bulk wine sellers, independent winemakers and merchants. Annual sales volume reaches 140 million bottles, with 74% sold in France (25% of which is sold directly to consumers in the vineyards), and the remaining 26% being exported. Belgium is the largest importer of Alsace wine by volume, followed by Germany and the United States (CIVA figures, internal report 2020).

Organic and biodynamic viticulture is a prominent and growing phenomenon in Alsace. It began in the 1970s and is currently practiced on 2,200 hectares of vineyards by 275 organic winegrowers, including 50 who follow biodynamic principles. Over the past decade, the area of organic vineyards has quadrupled, and the number of organic winegrowers has sextupled. Most Alsace wines are made from a single grape variety. Since the end of World War II, Alsace has been allowed to label its wines by grape variety in the Germanic tradition, as long as these wines consist of 100% of the indicated grape variety. Alsace also continues to bottle its wines in the traditional German style, using tall and fluted bottles known as "flûtes d'Alsace".

Pillar 1. The human dimension: the Alsace vineyard possesses a strong human aspect, exemplified by the vineyards' ability to build and nurture close relationships with consumers in various consumption contexts (wine route visits, cellar tours, tastings, etc.). Alsace is a reflection of the cooperative model, represented by several cooperative cellars that share common values, emphasizing the human element, as well as the principles of understanding and collaboration.

Pillar 2. Terroir: according to CIVA data, Alsace is the only region in the world that encompasses all possible terroirs. This significantly influences production methods (diversity in terms of practices) and the range of wines produced (diversity in terms of offerings). It is worth noting that Alsace wines are influenced by a double culture: Latin (Roman) and Germanic (or more accurately, Alemannic).

Pillar 3. Food and wine pairings: the third characteristic relates to the versatile and multifaceted nature of Alsace wines. These are adaptable to all contexts and consumption occasions (aperitifs, meal pairings, desserts) and can complement a wide range of dishes and world cuisines (savory, sweet, umami).

Box 12.2. *Three value pillars*

Specifically, wines with the AOC Vins d'Alsace or Alsace designation can only be transported, presented, offered for sale or sold in bottles, in accordance with regulations in effect since 1972[11]. Therefore, wines labeled AOC Alsace must be bottled in the same region where they are produced. In terms of marketing practices,

11 See: https://www.legifrance.gouv.fr/loda/id/JORFTEXT000000864835/ [Accessed September 15, 2022].

the communication surrounding AOC Vins d'Alsace, managed by CIVA, employs techniques similar to those used for brand promotion: a distinct visual identity, a dedicated website, a logo[12], storytelling (historical narratives, testimonials, life stories) and an identity built upon three core values[13] (see Box 12.2).

Therefore, the Vins d'Alsace brand serves as an umbrella brand for the promotion of all wines in the Alsace region, making it a case of a regional brand. We aim to explore the image it conveys to Alsatian wine producers to identify potential synergies with actors like the CSE in promoting wine and wine tourism in the region. Case study 2 is designed to address this objective.

12.3.2. *Data collection*

Our methodology consists of two case studies (corresponding to our first two research objectives) presented in Table 12.1.

	Study 1 Image of the CSE	Study 2 Image of "Vins d'Alsace"
Study objective	Analyzing the CSE's brand image	Analyzing the brand image of AOC Vins d'Alsace
Collection method	Qualitative study: semi-structured interviews	Qualitative study: semi-structured interviews
Sample	18 respondents	8 respondents
Respondents' profile	– 11 wine professionals (73% men, 64% winemakers, 90% members of wine brotherhoods including 40% of the CSE) – 7 consumers (57% women, 3 former customers, 4 potential customers)	8 Alsatian wine producers (62% men, 4 winemakers, 3 winegrowers, 1 cooperative winery)
Data collection tools	– Interview guide – Manual recording and transcription of interviews – Average interview duration: 1 hour	– Interview guide – Manual recording and transcription of interviews – Average interview time: 45 minutes
Data analysis	Manual thematic analysis	Manual thematic analysis

Table 12.1. *Research methodology: case studies 1 and 2*

12 The CIVA has created a protected Vins d'Alsace logo.

13 Pillars of values developed from marketing studies carried out by the CIVA in 2017.

12.4. Results

12.4.1. *Case study 1: CSE brand image*

The results are categorized by themes and the profiles surveyed, which include both members and non-members of the CSE among wine professionals. These results are discussed and supported by excerpts from interviews.

12.4.1.1. *Wine professionals' perception of the CSE*

The CSE can proudly claim a *strong reputation* in the region ("Personally, it's the only one I know in Alsace", "the most well-known, the Saint-Étienne brotherhood, truly the most significant, and I believe the oldest"). It is *associated with clear, distinct and positive attributes* ("the oldest", "the most important", "the most prestigious", "one of the most renowned in Alsace"). Furthermore, it is recognized for its *perceived usefulness* in promoting, representing and introducing Alsace wines. The analysis of how wine professionals perceive the CSE reveals specific and favorable associations that the CSE can leverage to refine and clarify its brand identity. It is worth noting that perceived usefulness varies, depending on whether an individual is a member or non-member, and these differences will be discussed below.

12.4.1.2. *Perception of the CSE by its members*

Members have positive and generally consistent perceptions of the CSE, whether in terms of *values* (e.g. tradition, conviviality, family, transmission) or *symbols* (e.g. red suits with white gloves, Sigille, the castle, wine library). The *perceived usefulness* of the CSE operates on two levels: *externally*, where the CSE plays a role in promoting the collective, meaning all Alsatian wines, and *internally*, for the members, where the CSE provides a space for conviviality, transmission, meetings and exchanges around a common passion. It also serves as a support network and an educational framework for training ("comparing our current methods, our current ways of doing things with what was done before, in the past", "we learn from each other about their knowledge of wine"). However, although the *sense of belonging* generates pride and recognition among members (e.g. feeling fortunate, privileged, gaining recognition, leaving a mark in history through the preservation of bottles in the wine library), it does not drive their *commitment* and *involvement* in CSE activities. This is a significant hindrance to its development.

12.4.1.3. *Perception of the CSE by non-members*

In contrast to members, non-members exhibit negative associations with the CSE, which may partially account for their non-membership. These associations revolve around four interconnected deficits: *lack of openness* towards young people, women and others in general ("We don't like the somewhat exclusive nature that the brotherhood can project to people", "I think it's essential to rejuvenate the Grand Council and diversify the profiles by including women and younger individuals"), *lack of modernity* (aging, austerity), *lack of information* about membership and its functioning ("That's why I've never attempted to join such brotherhoods because I've always believed you needed sponsorship", "It's quite obscure how it operates") and the *lack of perceived usefulness* ("We don't see the benefits it could provide us in the short, medium, and long term"). These findings collectively highlight areas for potential improvement for the CSE, when targeting this specific group of non-members.

12.4.1.4. *Consumers' perception of the CSE*

Clients and potential clients (outside the wine sector) of the CSE's services assessed the image of the CSE in two parts, its overall image as an institution and historical place and as a service provider (castle rental, various workshops, etc.).

12.4.1.4.1. Overall Image of the CSE

Consumer associations with the CSE can be grouped into three main categories: the *medieval aspect* ("being old-fashioned, reminiscent of illuminated manuscripts, bearing a crest or seal, and having an overall somewhat antiquated appearance"); the *connection to the Alsatian vineyard, heritage* and *reputation* (linked to a passion for wine, terroir, history, tradition and the broader Alsatian and wine heritage, as well as its association with a renowned place, "the Saint-Étienne brotherhood is a place of renown, prestige of Alsatian wines"; "I think of the castle, I think prestige, tradition"; "Well the setting, eh, magnificent, the services around wine, the geographical position") and the *lack of openness* (seen as a rather closed-off community, with suggestions "to provide more information about members and their activities to make it less exclusive"). Consequently, the image of the CSE is a blend of positive associations, emphasizing prestige and the preservation of traditions and heritage, and negative associations, primarily rooted in its medieval, visual and aesthetic identity and the perception of being a closed and exclusive group. These consumer associations align with the perceptions gathered from wine professionals.

12.4.1.4.2. Services offered by the CSE

For customers outside of the wine sector, the services provided by the CSE are linked to a *positive reputation and quality*. They appreciate the flexibility and attentiveness of the CSE, with remarks such as "they were flexible and attentive, going out of their way to accommodate us" and recognition of strengths like accessibility and prestige. However, there are some negative associations due to the *lack of information* on the services ("I didn't know that I could rent a room in the castle"), and the perceived level of after-sales service is low, with comments like "I received the program about a month ago *via* email, and, as much as I dislike receiving many emails, the brotherhood could improve its communication".

In summary, the CSE maintains positive associations, regardless of the profile surveyed. These associations relate to the tradition it upholds, the heritage it represents, the values it promotes and its utility to various stakeholders, including wine professionals and customers. In addition, the CSE has negative associations among non-members (members do not mention negative aspects). These negative associations are related to the lack of openness, the perception of elitism or exclusivity, the perceived lack of modernity due to its medieval and old-fashioned image and the overall lack of clarity and information.

12.4.2. *Case study 2: Alsace wines brand image*

The results of case study 2 are organized according to five themes. Within each theme, several sub-themes are discussed and illustrated with excerpts from interviews.

12.4.2.1. *Attributes and values*

The attributes of Alsace wines can be summarized along five axes: the *terroir* ("is not a factory, there is a real precision of the terroir"), *conviviality* ("very convivial"), *diversity* ("what is interesting is to have a diversity in the grape varieties. These grape varieties are also known abroad"), *human* ("what is quite important is tradition, history … we talk about it, it is an added value for us, the human side") and *environmental protection* ("the environmental side is very important here, the vineyard is very green, sustainable development, we are happy to be at the forefront"). These aspects align with the three pillars of the AOC Vins d'Alsace, indicating a convergence between the projected identity and the perceived image, at least at the level of the target of wine professionals in Alsace. They are also congruent with CSE associations (terroir, conviviality). These crossing points therefore constitute potential levers for a synergy effect between these actors.

12.4.2.2. *Personality(ies)*

The personality of Alsace wines is perceived as *multifold* ("There is not a single personality for all wines, it is impossible. It's like for humans, they are complementary") with a strong dimension of *authenticity* ("wines of character … fairly frank wines. They have personality, they know where they are going … The wines are clearly identified"). It is nevertheless possible to synthesize all the characteristics in four dimensions as follows: *benevolence* (generous, friendly, likable), *cheerfulness* (cheerful, festive, dynamic), *authenticity* (honest, frank, true, transparent, characterful) and *uniqueness* (precise, influential, identifiable). These dimensions corroborate the three identity pillars developed for the AOC Vins d'Alsace, although there is an emerging dimension, that of gaiety. In addition, these dimensions diverge from the associations identified at the CSE level.

12.4.2.3. *Target: experts versus novices*

The target audience for Alsace wines is perceived as *diverse* ("there is something for every budget, you can find entry-level wines and then Grand Crus for more specific meals. There is no reason for Alsace wines to remain in this non-consumption of wine drinkers"). There is a distinction between *varietal wines* and *terroir wines*. The former are considered entry-level and intended for the most novice consumers. The latter are for experts ("it is clear that Alsace is for connoisseurs. The terroir ... Experts are fascinated by the wines of Alsace. The quality/price ratio is incredible, but when you are in front of a non-expert, it is too expensive") and can be "confusing" for novices ("more for experts, due to Alsace terroir wines. Less knowledgeable consumers are lost"). Due to their diversity, Alsace wines seem to cater to two opposing targets: experts and novices. This dual perspective on user imagery can potentially harm the brand image of these wines, as it leads to confusion. This mirrors the situation with CSE, where the user's image generates negative associations regarding CSE's image (not for everyone, not for young people, not for women, for a privileged target, for older individuals).

12.4.2.4. *Tradition versus innovation*

Tradition is considered an inherent association of Alsace wines, in addition to being one of their fundamental values. However, respondents emphasize that the region should prioritize innovation to make the perception of Alsace less traditional ("Innovation is important, but Alsace isn't quite there yet, even though a lot of work is being done. Alsace wines have remained somewhat traditional, consumed at the table, as an aperitif, but not all the time, and the consumption is quite targeted").

In this context, the labels and bottle shapes in Alsace are seen as too traditional ("You see, a slightly different bottle, even for Grands Crus with screw caps. Labels are often very traditional"). The primary drivers of innovation include *wine tourism* ("Wine tourism will establish a connection with the customer who returns home, and, if we've done our job well, they will remember our visit, in addition to the wine"), *digitization* through the use of social media ("The most obvious aspect is the use of social media"), and production tools perceived as essential for efficiently producing more precise and environmentally friendly wines ("We strive to incorporate very new tools and intelligent systems, enabling us to work optimally in small teams").

In summary, positive associations revolve around the quality (aromatics, gastronomy, vivacity, freshness), values (tradition, terroir, conviviality, humanity, environmental protection) and the personality (benevolence, cheerfulness, authenticity and uniqueness) of Alsace wines. In addition, wine tourism activities and innovations in terms of production tools (modern, precise, efficient) are seen as value-creating and highly rated. However, a significant negative aspect is the lack of innovation (bottle shapes, labeling, communication via social networks), and there is a strong emphasis on the need for modernizing the image.

12.5. Summary of results

In both case studies, paradoxes or dualities emerged from the results. Concerning the CSE, its utility is acknowledged; however, its operational methods and membership conditions remain unfamiliar or carry a priori negative connotations. It is perceived as a prestigious and welcoming institution, and yet at times it is described as obscure or enigmatic. In the case of AOC Vins d'Alsace, there is no consensus on elements requiring modernization, with some seeing the bottle format as a strong symbol and a source of uniqueness, while others view it as overly traditional. Lack of agreement also extends to the characteristics of Alsace wines, judged as unsuitable for all target groups, being labeled as too traditional, complex or expressive. The diversity of these wines is seen as both a strength, catering to various tastes and budgets, and a weakness, resulting in ambiguity and confusion, especially among novices, while primarily appealing to experts.

The same holds true for the perception of the country of origin of Alsace wines. On the one hand, the association with "made in France" is considered too weak, even confusing (with Germany or Austria), which hinders Alsace wines from benefiting, at least on an international scale, from the positive attributes typically linked to French wines (quality, prestigious image, gastronomy, reliability). On the other hand, this is also seen as a factor of differentiation (appreciated originality) that should be emphasized for Alsace wines. These paradoxes highlight an inconsistency in terms of image (reflection of identity) among the target audience, presenting marketing challenges.

First, the brand identity needs to be clearly formulated, representing the initial challenge. Second, the subsequent challenge is to effectively convey these identity elements to the target audience. Based on these findings, we will provide managerial recommendations in the following section.

12.6. Discussions and recommendations

This research focuses on the importance of creating synergies between a regional wine brand (Vins d'Alsace) and the branding of a wine tourism provider (the CSE), recognizing that wine tourism is a central theme in Alsace. The managerial implications of this research are directed to both the CIVA, responsible for promoting the Vins d'Alsace brand, and the Grand Council and management of the CSE. However, these implications may also apply to other stakeholders (such as tourist offices, winegrowers, etc.) and extend to other wine regions featuring AOCs and wine brotherhoods. Our recommendations are organized around four key areas to address our third research objective (identifying synergy levers) and provide solutions to the issues raised in our two case studies. These issues include: (1) the dual challenge of maintaining traditions while embracing modernization; (2) the paradox between the human and humble aspects and the need for greater openness and diversity; (3) addressing confusion related to target audiences; and (4) establishing a clear territorial identity for the Alsace image internationally.

12.6.1. *Tradition versus innovation*

There is a genuine attachment to ancestral heritage and a strong commitment to preserving and passing it on. However, whether in the case of wines or the CSE, this dedication to tradition is in conflict with a pressing need for rejuvenation (in terms of packaging, promotional tools, wine tourism, targeting methods, access to information, etc.) while still upholding core values (such as terroir and the human element). Tradition is appreciated and critiqued simultaneously. The need for modernization is considered essential to remain competitive and at the forefront, given that innovation is becoming the industry norm. In addition, even though the Alsatian vineyard is known for its humble identity, its proximity to people, conviviality and a culture of sharing, the paradoxical lack of openness to new ideas and innovations is seen as a major impediment to its development. While the CSE is more rooted in tradition and is perceived as more closed, Vins d'Alsace are characterized by dynamism, innovation, modern communication and digital tools aligned with current trends. Therefore, potential synergy effects exist between the CSE and Vins d'Alsace. As a result, we propose the following five recommendations:

– Utilize the CIVA's digital channels (official website, "Route des vins d'Alsace" website, official accounts on various social networks) to enhance the visibility of the CSE among potential consumers and tourists. This strategy would help reinforce the perception of openness and accessibility of the CSE, especially by showcasing event offers as exceptional opportunities to explore Alsace wines. In addition, the CIVA could benefit from integrating the heritage dimension into its digital communication.

– Pool human resources, especially through the CIVA's marketing services, which could assist the CSE in fostering communities of Alsace wine enthusiasts.

– Establish a Cité des Vins d'Alsace that blends the modernity of Alsace wines with the tradition upheld by the CSE. Unlike other French wine regions that intentionally create innovative *Cités des Vins*, Alsace should consider a distinctive approach, leveraging its existing assets in authentic wine tourism and tradition.

– Emphasize the human element by highlighting the role of the Grand Master as a genuine ambassador for Alsace wines, supported by an official status justifying compensation. This way, the role of a Grand Master would be officially recognized as a lever for peer recognition and a potential source of recognition for the younger advisors of the CSE. To enhance the synergy regarding the human aspect, the life stories of various members (diversifying their ranks) of the CSE, similar to the life stories of winemakers, could be featured on the CIVA website.

– Re-evaluate governance by suggesting connections between the CIVA Board of Directors and the CSE Grand Council. This would entail collaboratively developing communication strategies, among other things.

12.6.2. *Prestige and heritage*

The CSE possesses a substantial heritage potential and legacy, including prestige through the castle and its associated events, a historical connection to medieval traditions and a sense of cohesion and duty, tangible heritage through the wine library and intangible heritage via uniforms, ranks, vineyards, rituals and other elements of folklore. These elements are valuable assets for enhancing the richness of the Alsace wine experience with strong and memorable symbols, as well as cultural discoveries.

As such, we identify potential synergy effects between the CSE and Vins d'Alsace, and make the following recommendation: the Vins d'Alsace brand primarily emphasizes the "wine" product, while the CSE develops offers for experiencing the "wine" product. The CSE's castle serves not only as the headquarters of a wine brotherhood but also as a unique venue for discovering Alsace wines. The CSE is a historic institution that visitors can explore to immerse themselves in Alsatian culture and to discover the region's primary agricultural product, wine, through visits to the wine museum and wine library. In this sense, the CSE serves as an entry point for introducing Alsace wines to a target of customers/tourists who may not have initially visited for wine. It contributes to the promotion of the destination. Therefore, it is essential to reconsider the customer/tourist journey based on their profile, in order to intersect the paths between regional tourism and discovering regional wines within an experiential context.

12.6.3. *Diversity of wines and targets: a strength or a weakness?*

The diversity of Alsace wines (grape vs. terroir) and, consequently, their target audiences (novices vs. experts) is a characteristic that lacks consensus. There is no unanimous agreement on whether it is a strength (originality) or a weakness (a source of confusion). This leads to ambiguity regarding the positioning of Alsace wines in France and internationally. On one hand, the CSE targets an expert audience (organizing events, workshops and meetings for enthusiasts) and values the diversity of wines in its wine library. However, with a variety of offers and packages, it also provides initiation for novices, even though they are not the primary target. On the other hand, the CIVA's role is to make Alsace wines accessible and understandable to the majority of consumers (*Alsace en Terrasse* and *Alsace Rocks*[14], Alsace Wine Fair, etc.) and to offer expert guidance (importers, press, etc.). Both have the potential to reach all target audiences through customized offerings tailored to their expectations and preferences. Therefore, we observe a convergence that can serve as a basis for synergy between the CSE and Alsace Wines, leading to the following recommendation: create different online customer/tourist journeys based on their level of expertise, ranging from "which Alsace wine interests me" to "what CSE activity can enhance and enrich this interest". Ideally, specific packages would link preferred wines with the desired experience:

– For expert profiles, for instance, if there is an interest in terroir wines, offer a "vintage wines" workshop at the CSE.

– For novice or beginner profiles, emphasize the "initiation" and "discovery" workshops focused on wines that are easier to taste.

12.6.4. *International*

From an international perspective, our results highlight the weak connection between Alsace wines and the "made in France" label. It is perceived as both a strength and a weakness. The question then arises as to whether this association should be strengthened or simply set aside.

14 Communication campaigns developed by the CIVA in 2020.

In reality, there is a genuine need for consistency in terms of the image to be projected, both nationally and internationally. We therefore recommend developing *unifying projects* that enhance the region's appeal and position Alsace not as the third most visited wine region by tourists in France but as the leading wine destination. However, one indispensable condition is the ability of various stakeholders to work cohesively towards this common goal.

Figure 12.2 represents the integrative model of the research which illustrates our results as well as our recommendations in terms of synergy effects to be developed.

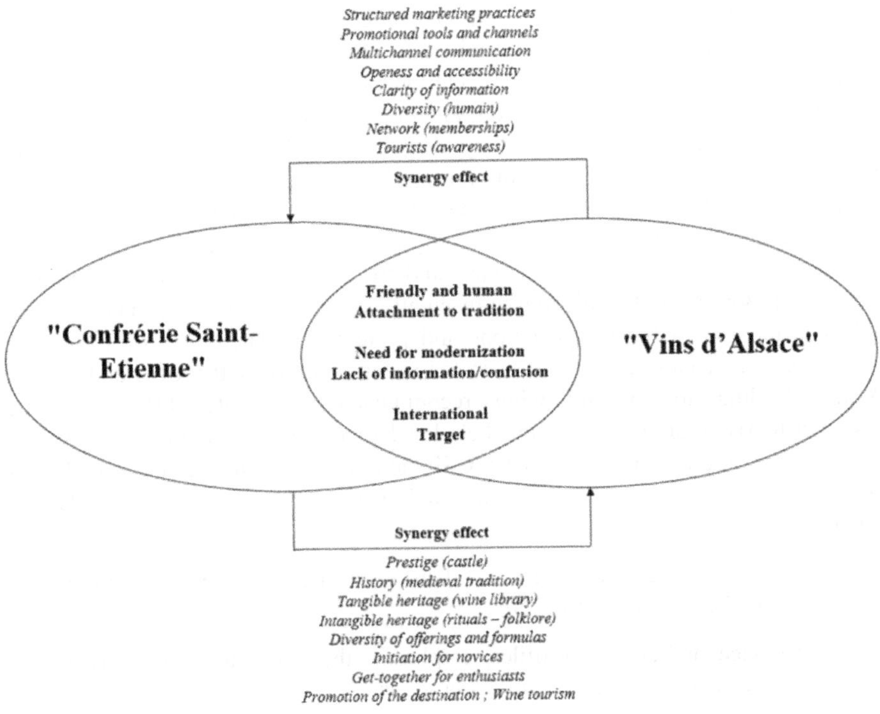

Figure 12.2. *Integrative model of synergy effects between the two brands, based on the results of the two case studies*

12.7. Conclusion

Not religious or mystical, and certainly not secret societies, wine brotherhoods continuously pique curiosity, especially due to the hierarchical titles given to their respective council members. However, they still play a significant role in promoting wine tourism in wine regions and should not operate in isolation.

The AOC Vins d'Alsace is facing a dilemma: image and positioning consistency (tradition vs. modernization), geographical anchoring and target audience. This work has illuminated the strengths and areas for improvement of these two regional actors in the promotion of wine and wine tourism through two case studies. While each possesses unique characteristics, both encounter similar challenges and should work together synergistically.

12.8. Acknowledgments

The authors would like to thank their students for contributing to the two case studies:

– For case study 1, we thank Naïm Chaabane, Juliette Blouch-Cosquer, Audrey Boete, Pauline Deshayes, Nathan Deubel, Steve Dita Leta, Lola Martin, Pierre Quillien, Cheyenne Tyll and Marion Verdin (Alumni of the Grande École Program, specialization in "Co-op Marketing" 2021/2022, EM Strasbourg).

– For case study 2, we thank Gracie Bolland (Alumnus Master 2 Management and Business Administration (MAE) Tourism Management 2019/2020, EM Strasbourg; member of the "Wine & Tourism" corporate chair).

12.9. References

Aaker, D.A. (2012). *Building Strong Brands*. Simon and Schuster, New York.

Boland, G. (2020). Regional wine brand equity: The case of alsace wines in the US market. Research Thesis, EM Strasbourg/Université de Strasbourg, Strasbourg.

Charters, S. and Spielmann, N. (2014). Characteristics of strong territorial brands: The case of champagne. *Journal of Business Research*, 67(7), 1461–1467.

Easingwood, C., Lockshin, L., Spawton, A. (2011). The drivers of wine regionality. *Journal of Wine Research*, 22(1), 19–33.

Frochot, I. (2003). The role of wine brotherhoods in the development of French wine Tourism. *Journal of Tourism Studies*, 4(2), 29–27.

Garrier, G. (1999), Les confréries autrefois. *Actes du Colloque Mondial de la Commanderie des Costes du Rhône*, Université du Vin, Avignon.

House, M. and Bruwer, J. (2003). Has the era of regional branding arrived for the Australian wine industry? Some perspectives. *Australian and New Zealand Grapegrower and Winemaker*, 479, 56–61.

Kavaratzis, M. and Hatch, M.J. (2013). The dynamics of place brands: An identity-based approach to place branding theory. *Marketing Theory*, 13(1), 69–86.

Lopes, C., Leitão, J.R., Engifo-Gallego, J. (2018). Place branding: Revealing the neglected role of agro food products. *International Review on Public and Nonprofit Marketing*, 15(4), 497–530.

Menival, D. and Charters, S. (2014). The impact of geographic reputation on the value created in Champagne. *Australian Journal of Agricultural and Resource Economics*, 58(2), 171–184.

Ruzzier, M.K. and de Chernatony, L. (2013). Developing and applying a place brand identity model: The case of Slovenia. *Journal of Business Research*, 66(1), 45–52.

Thode, S.F. and Maskulka, J.M. (1998) Place-based marketing strategies, brand equity and vineyard valuation. *Journal of Product & Brand Management*, 7(5), 379–399.

PART 4

International Context

13

Perception of Imported Wines by African Consumers: Where do Bordeaux Wines Stand?

13.1. Introduction

France is still the second-largest consumer of wine in the world, behind the United States. However, this context is marked by reduced consumption (CNIV 2019). While professionals are striving to revitalize wine consumption in France through the development of new occasions, such as aperitifs, they are becoming increasingly reliant on exports for their survival. 80% of these exports are directed to around ten countries in Europe, North America and China.

French wine exports to Africa still account for a marginal volume (FranceAgriMer 2021). However, Cameroon and Ivory Coast are emerging as significant destinations for Bordeaux wines. In 2021, Ivory Coast imported 4 million bottles of Bordeaux, valued at 9 million euros (a 126% increase in volume and a 107% increase in value compared to 2020). Thanks to this, this African country made it into the top ten primary export destinations for Bordeaux wines. During the same period, Cameroon experienced a 34% increase in the volume of Bordeaux wine imports and a 64% increase in value, ranking twelfth among the leading importers of Bordeaux wines globally (Baudouin 2022).

Moreover, the aggregation of the fifty-four African countries represents a substantial market potential and thus a significant opportunity for wine producers. Wine consumption in Africa has markedly increased over the last decade and

---

Chapter written by Mathieu NLEND-MOMNOUGUI, Jean-Philippe GALAN and Franck DUQUESNOIS.

currently constitutes 5 to 6% of global wine consumption in volume. According to Kassous (2019), wine consumption in Africa had already reached nearly 1 billion liters in 2017, compared to 865 million liters in 2013 and 736 million liters in 2009. Since then, African wine consumption has been trending to surpass the total wine consumption of Spain, which is the seventh largest wine-consuming country globally, with a volume of 1 billion and 30 million liters of wine consumed in 2017 (Wine Institute 2020).

The prosperity and stability of several African countries are gradually attracting players in the wine industry, increasing competition on the continent (Duquesnois et al. 2021). Over the past 15 years, New World wines appear to be gaining ground in Africa, similar to wines from South Africa, Chile or California. In this context, how can French wine establish a long-term presence in Africa?

Currently, French producers export their products to the African market using the packaging and graphics from France. The same applies to communication. The cultural aspects of wine consumption in Africa are still underexplored in the academic literature. Therefore, it would be interesting to comprehend how Africans evaluate wine quality to reduce risks in their purchasing decisions. This research represents the initial step in the cultural analysis of African consumer responses to the marketing strategies of Bordeaux wine producers. While France's international wine market is becoming more premium, AOPs account for 38% of exports, with the Bordeaux appellation being the most exported (FranceAgriMer 2021). The aim here is to understand consumption behavior in various contexts where Africans buy Bordeaux wine based on four main purchase intentions: personal and individual consumption (at home), personal and public consumption (restaurant/bar), impersonal consumption (gift) and festive consumption (celebrations of socio-cultural events).

This leads us to a series of questions. What social and taste factors lead to the consumption of French wines in the African context? How do Africans assess the extrinsic and intrinsic indicators of French wines to minimize the risk of a bad purchase? Which informational and aesthetic elements of French wine packaging have the greatest influence on African consumers during the purchase process?

In the first phase, we will conduct an initial literature review on knowledge related to wine consumption in Africa, as well as on the foundations of wine consumer behavior. In the second phase, we will proceed with a preliminary qualitative study, using a correspondence analysis, which will help us better understand the distinctive position that French wine and Bordeaux wine, in particular, hold among African consumers.

13.2. Attitude of African consumers and wine market in Africa

13.2.1. *Attitude of African consumers*

Marketing literature links consumer attitudes to their beliefs and relationships with specific products or services within a given environment. These predispositions serve as filters through which consumers evaluate the products presented to them. In the African context, numerous studies have explored consumer attitudes toward imported products or services. Most of these studies highlight the impact of specific attitudes on the preferences of African consumers. For instance, Quaye et al. (2009) discovered that socio-cultural factors, including ethnic dietary traditions and habits, influence the attitudes of Ghanaian consumers towards certain types of food products. Regarding services, Dadzie et al. (2003) found that social norms and beliefs influence the long-term saving behavior of rural households in Ghana. In the realm of services, Gait and Worthington (2015) in Libya, along with Ltifi et al. (2016) in Tunisia, identified the strong influence of religious guidelines and practices, such as Sharia, on customer attitudes towards retail banks in these two countries. Among the influencing attitudes, the health consciousness of urban African consumers also appears to impact purchasing decisions for products. For example, in Kenya, urban consumers were willing to pay a premium for new drinking water filters due to the perceived health benefits (Brouwer et al. 2015). Similarly, Ghanaian consumers were willing to pay more for vegetables produced using treated wastewater compared to untreated wastewater in urban and peri-urban vegetable production (Yahaya et al. 2015).

Furthermore, environmental concerns have also been revealed as decisive factors in the purchasing decisions of African consumers. Dubihlela and Ngxukumeshe (2016) discovered a significant association between the attributes of eco-friendly retail products and their purchase by South African consumers. Likewise, Thanika et al. (2012) observed that the majority of Mauritian consumers exhibited a high level of interest in environmental protection, which seems to influence their decisions to buy green products.

Another highly influential attitude in the African market is the "country factor". This refers to the positive or negative impact that the country of manufacture of a product can have on a consumer's perception. Studies on the "country factor" are based on two principles:

– The psychosocial construction of ethnocentrism: this refers to the belief that consumers in a given country believe that products and services from their country are superior to those produced in other countries. John and Brady (2011) found that Mozambican consumers were moderately ethnocentric and had relatively unfavorable attitudes towards South African products.

– The country of origin effect: this describes the positive or negative image of products originating in a given country among consumers in the host country. Previous studies on the country-of-origin effect in some emerging economies suggest that products made in advanced Western countries are perceived to be better than those in developing or less developed countries. This applies even more in the case of wine, as few African countries produce it. However, some results seem to contradict this view with regard to price. For example, Saffu and Walker (2006) found that Ghanaian consumers are willing to purchase best value products regardless of country of origin. Similarly, in Uganda, Ngoma and Ntale (2015) found, regarding the choice of pharmaceutical products, that affordability of the price as well as the perceived quality of the brand remained stronger determinants than the "country of origin" determinant. These conflicting results would explain why, although Bordeaux wines enjoy a strong "country of origin" effect in Africa, Spanish wines that are more affordable in terms of price remain the most consumed.

In addition to aspects related to the attitudes of African consumers, the influence of perceived quality on the purchasing decisions of African consumers has been the subject of several researches. In Tanzania, for example, consumers were willing to pay relatively higher prices for better-quality condoms (Brent 2009).

13.2.2. *Wine market in Africa*

Africans typically use alcoholic beverages and wine, especially for celebrating significant rites and social occasions like marriage, funerals or widowhood (Nugent 2021). Paradoxically, despite Africans' strong religious attachments (Mbiti 1990), some African societies resist the consumption of alcohol in all forms. For instance, complete alcohol abstinence is a crucial identity marker among Sub-Saharan African Muslims, where there is a 30% Islamization rate. In regions like Senegambia, conversion to Islam was traditionally marked by the permanent renunciation of palm wine (Nugent 2021). Additionally, a significant proportion of Nigerian Muslims abstain from alcohol, including wine (Brouté 2016). Conversely, the African market has long been dominated by entry-level wines, with an abundance of Spanish wines (e.g. Tsapi et al. 2009). However, the prosperity of certain oil-producing African countries, such as Nigeria and Angola, has given rise to a wealthier middle class with distinct consumption habits, favoring ostentatious consumer products like wines and spirits over beer (Darbon 2012). In Nigeria, wine consumption has shifted towards not only the wealthier social classes (Ogbeide and Ele 2015) but also the upper middle class, whose evolution has been a key factor in the market (Jacquemot 2012).

13.2.3. *Intent to purchase*

Customer satisfaction after a purchase depends on the product's performance relative to the customer's expectations. Therefore, if the product's performance falls short of these expectations, the customer is dissatisfied. Conversely, if the product's performance aligns with consumer expectations, the customer is satisfied. In addition, some authors point out that customer expectations are shaped by communication promises, advice from friends, previous shopping experiences, and expert opinions. Given this, understanding the purchasing intentions of Africans is crucial in determining their expectations regarding Bordeaux wines.

Wine is a product that is predominantly purchased in anticipated social contexts (Hirche and Bruwer 2014). Whether for personal and individual consumption (at home), personal and public consumption (restaurant/bar), impersonal consumption (gift) or festive consumption (socio-cultural event celebrations), the choice of wine is influenced by the specific situation. Hall and Lockshin (2000) emphasized that the consumption context plays a role in determining the importance of various product attributes and the final selection of wine. For instance, consumers may opt for higher-priced wines for special occasions like dinners or business meetings, while lower-priced options are preferred for more informal or personal events (Charters 2006). In the African context, Bordeaux wines are typically associated with grand occasions and prestigious gifts.

13.2.4. *Influence of extrinsic indicators*

13.2.4.1. *Country of origin*

Martinez-Carrasco et al. (2005) elaborated on the significance of the country of origin in assessing wine quality, particularly when consumers lack experience and other more objective evaluation methods are not available. This perspective appears to be well suited for Africa, where consumers often do not have access to objective wine evaluation tools at the time of purchase, leading them to be influenced by the wine's country of origin in their decision-making.

Chaney (2002) observed that large retailers tend to organize wines on their shelves by their countries of origin, further highlighting the importance of this criterion in choice. Additionally, according to Perrouty et al. (2006), when the wine's region of origin is associated with a certain level of prestige, as is the case with Bordeaux wines, any quality shortcomings are perceived less and are more readily forgiven.

13.2.4.2. *Packaging*

Behaeghel (1991) asserted that packaging could be the most significant communication medium, as it influences nearly all consumer categories and is present at the crucial moment when the purchase decision is made.

Aesthetic and textual elements are among the fundamental elements used by African consumers in purchasing situations. In countries with a strong wine culture, grape variety, brand, and price have been identified as the key elements used by consumers in their purchasing decisions (Thomas and Pickering 2003). Experienced consumers choose wine based on their personal experiences and the information provided by various elements of the packaging, while others may rely solely on the label to meet their information needs. In contrast, less wine-savvy consumers may place great importance on the country or region of origin (AOP) of the wine when evaluating its quality before making a purchase. Today, there is also a growing interest among African consumers in wines with a high alcohol content. The symbolic elements used on wine packaging are interpreted by consumers through verbal-pictorial metaphors (Caballero 2009). Wine labels often feature metaphors like castles, vineyards and creatures, such as the kangaroo on the Yellow Tail[1] label (Lehrer 2009). Novixoxo et al. (2019) note that significant attention is given to the shape, size, label placement and the inscriptions present on wine labels. However, the overall meaning of a visual message composed of symbols and colors is influenced by the cultural and sociocultural context of the receiver (Joly 2005). In this regard, Tsapi et al. (2009) assert that Cameroonian consumers associate quality with wines in bottles with hollow bottoms.

13.2.4.3. *Subjective knowledge*

Subjective knowledge is frequently linked to extrinsic indicators and is defined as the level of expertise perceived by the consumer, gained through familiarity with a product category. It typically relies on personal experience, exposure to marketing information and recommendations.

Numerous pieces of evidence support the impact of extrinsic cues when consumers refer to subjective knowledge rather than objective knowledge. Unlike consumers with a high level of objective knowledge, those who rely on subjective

1 Yellow Tail is an Australian wine brand launched in 2001 by the Casella family that relied on an original marketing positioning characterized by a kangaroo on the label. The brand would present itself as a "fun" brand, easy to access and aimed at new wine consumers (source: Strategy and Leadership, Value pioneering – how to discover your own blue ocean interview with W. Chan Kim and Renée Mauborgne).

knowledge do not have an extensive array of relevant information to draw upon. They can typically only recall a few brand names and perhaps just one or two specific attributes for each of them. These individuals filter out the attributes that they do not comprehend and utilize extrinsic indicators like brand, price or country of origin to compensate for their lack of information.

13.2.4.4. *Objective knowledge*

Objective knowledge refers to precise information stored in an individual's long-term memory. This type of knowledge is primarily based on cognitive learning and our experiences with the product category being assessed. It has been observed that consumers with high levels of objective knowledge can more easily and accurately discern the essential attributes of products and services, while overlooking less crucial product features when making decisions, whether these features are intrinsic or extrinsic. An expert possesses knowledge rooted in extensive personal experience of objective knowledge. As a result, they are less likely to seek advice from external sources and do not place great trust in product-advertised benefits during information gathering. It has also been established that experts primarily focus on intrinsic cues and may not consider factors like brand, price or country of origin. However, extrinsic indicators are not dismissed if they genuinely predict quality. Even a true expert can take biases into account when these biases significantly affect a particular type of product, as is the case with the country of origin, which can indeed predict product quality compared to other countries. Furthermore, judgment mechanisms can be quite significant for a specific group of consumers. According to Favier et al. (2018), there are five major categories of wine evaluation mechanisms: appellations (labels, certifications, etc.), cicerones (critics, experts, guides), rankings (awards, accolades, etc.), confluences (customer service skills and information at points of sale) and networks (interpersonal relationships that facilitate the exchange of knowledge). These judgment mechanisms are crucial for "intermediate" consumers, those who aim to apply objective knowledge in purchasing situations.

13.2.4.5. *Reminder: intrinsic indices*

It is important to note that an intrinsic indicator is defined as any attribute inherent to the primary function of a product, such as a car engine's capacity or the taste of a beverage. According to the literature, intrinsic indicators are more credible than extrinsic indicators, but this is not the case when consumers do not find them sufficiently predictive or lack confidence in their ability to assess them. Consumers are not always able to accurately evaluate intrinsic indicators before making a purchase, often due to a lack of understanding or inaccessibility. In the specific context of wine, previously experienced intrinsic indicators play a crucial role in repeat purchase situations.

13.2.4.6. *Impact of packaging in a purchasing situation*

13.2.4.6.1. Packaging

Packaging is defined as a combination of structural, graphical and verbal design features that can serve as cues for the consumer (Magnier and Crié 2015). Its verbal features encompass explicit textual information available on the packaging, usually providing consumers with details such as the brand name, flavor, nutritional content and expiration date. Packaging may be the most critical means of communication, as it influences nearly all consumer categories and is present at the pivotal moment when purchasing decisions are made. Its role is evident in the aesthetic assessment of a product (Celhay and Trinquecoste 2015), price and quality prediction, and, to some extent, expectations regarding taste evaluations.

The utilization of symbols is a significant practice in enhancing packaging. The symbolic content employed is expected to be interpreted by consumers through what is known as a verbopictorial metaphor. In the context of wine, the use of a label rich in symbolic content is a strategy for enhancing the wine's image, with verbopictorial metaphors like the images of castles on French (and pseudo-French) wine labels, or the previously mentioned "creatures" such as the kangaroo on the Yellow Tail label (Lehrer 2009). This adds value to the packaging.

In this regard, it is observed that there is currently a significant focus on the shape, size, and position of the label, as well as the information displayed on it, as it determines the level of consumer interest in the product. Concerning the label, the front portion piques consumer interest and serves the marketing function of attractiveness, while the back label contains technical product information.

13.2.4.6.2. Purchasing situation

Wine holds symbolic meanings linked to lifestyles, which can complicate the buyer's decision-making process. The significance of situational aspects in consumer behavior has been acknowledged by various research studies. Understanding the effects of the purchase situation in conjunction with consumer behavior forms a crucial foundation for tailored marketing efforts. The literature recognizes three types of situations that influence consumer behavior: (1) the communication situation; (2) the purchase situation; and (3) the consumption situation. While the purchase situation impacts brand, variety and form choices, the consumption situation is believed to affect the consumer's decision-making structure within the buying process. Furthermore, a consumer's involvement in the purchase and motivation to select the right product depend on both the purchase situation itself (e.g. urgency) and the consumption situation (e.g. catering to children's food).

13.2.4.6.3. Consumption situation

The concept of the consumption situation is regarded as significant in comprehending variations in consumer behavior. The consumption situation contributes to the significance of various product attributes and the ultimate choice of wine. For instance, a consumer is willing to pay a higher price when purchasing wine for a special occasion, such as a family celebration dinner or a crucial business meeting, whereas lower prices are more appealing to consumers buying wine for a less formal occasion.

13.2.4.6.4. Planned purchases

Wine is a product that is often purchased based on situational factors, and it is frequently bought in anticipation of social occasions. Wine buying patterns in Africa align with this and result in high consumption during celebratory events. Indeed, several studies indicate that wine consumption increases during festive moments. It is noticeable that gatherings in Africa, such as weddings, funerals, and tribal or political meetings, are increasingly associated with wine consumption, whereas in the past, these events were linked to traditional beverages like raffia or palm wines, corn-based alcoholic drinks and beer. Therefore, a significant consumption of wine in Africa is part of a planned purchasing process. It is a rational decision-making process that involves a longer period of product evaluation and a greater search for information than during unplanned purchases.

13.3. Exploratory study: perception of the international wine supply by African consumers

Before delving deeper into our research, it is important for us to examine how African consumers perceive wines from various countries that export to the African continent. To accomplish this, we conducted a qualitative analysis involving two samples of African consumers hailing from four countries, among the top five wine-consuming nations in Africa in 2021: Benin (5th), Cameroon (2nd), Côte d'Ivoire (3rd) and Togo (1st).

13.3.1. *Research method*

We conducted an elicitation of attributes associated with wines based on knowledge a priori (no tasting or information provided to the consumer) using the direct quotation method (Vernette 1994) with an initial sample of 49 African consumers. Initially, respondents were categorized based on socio-demographic characteristics: country, gender and the frequency of red wine consumption (see Table 13.1). Subsequently, they were asked to spontaneously express their thoughts on attributes related to French wines (excluding Bordeaux), Bordeaux wines, Portuguese wines, Spanish wines, South African wines, Californian wines and Chilean wines. Respondents who could no longer provide new attributes for a particular wine evaluated the next one. The interview grid employed followed the framework proposed by Vernette (1994). Following this data collection, lemmatization was performed, primarily involving the consolidation of terms in singular/plural, masculine/feminine and synonyms. This process resulted in a list of 16 attributes of interest (Vernette 1994), as presented in Table 13.2.

In the next phase, 47 African consumers were prompted to select attributes from a predefined list containing these 16 attributes identified earlier. Their task was to define their perception of seven imported wines: French wine (excluding Bordeaux), Bordeaux wine, Portuguese wine, Spanish wine, South African wine, Californian wine and Chilean wine. In total, 1,294 wine/attribute pairs were cited. Based on this data, a contingency table was constructed, which links the seven evaluated wines with the frequencies of occurrence of the 16 attributes for each of these wines (see the appendix in section 13.5). The analysis of this contingency table suggests that the distribution of attributes across the wines is not random ($\chi^2[90] = 158{,}278$; $p < 000$), and the correspondence factor analysis indicates relatively low inertia (0.122), indicating the limited exclusivity of wine/attribute pairs. The retained axes account for 73.9% of the total inertia.

Variables	Respondents
Gender	
Male	32
Female	17
Country	
Cameroon	39
Togo	4
Côte d'Ivoire	3
Benin	3
Frequency of red wine consumption	
Very rarely	4
Only on special occasions	11
From time to time	21
Quite frequently	5
Very frequently	8

Table 13.1. *Socio-demographic characteristics of respondents*

Attributes
1 Good wine
2 Wine quality
3 Poor quality
4 Soft wine
5 Strong wine
6 High price
7 Affordable price
8 Pleasant
9 Fruity
10 Prestigious
11 Mundane
12 Blending wine
13 Everyday
14 For moments with friends
15 For special occasions
16 Suitable for gifting

Table 13.2. *List of attributes of interest*

13.3.2. *Results*

As evident from the table provided in the appendix in section 135, Bordeaux wine has generated the highest number of pairs, suggesting that it is the most well-known and/or has the most enriched associated imagery. Predictably, attributes related to quality and price were the most frequently mentioned.

Figure 13.1 illustrates the inherent perceptual space of the evaluated wines, which was derived from a correspondence analysis. This method offers the advantage of projecting the rows and columns of the contingency table onto a graph, allowing for a simultaneous representation of the wine positions and attribute positions. Consequently, the analysis is highly intuitive and provides insights into three key aspects: (1) the relative positions of wines in relation to each other; (2) the relative positions of attributes in relation to each other; and (3) the positioning of wines in relation to attributes. Unlike a principal component analysis, where axes are defined by the contribution of each variable to component construction, the direction of the axes in this analysis is determined by the positions of wines and attributes, and can be inferred from observable contrasts. Attributes or wines closer to the center of the graph are considered less distinctive. Specifically, a wine at the center of the graph signifies that it does not exhibit a distinct profile compared to other wines concerning attributes. Similarly, an attribute at the center of the graph has not been differentially associated with various wines but has been assigned to most. Consequently, the points farthest from the center provide the most meaningful insights in the analysis.

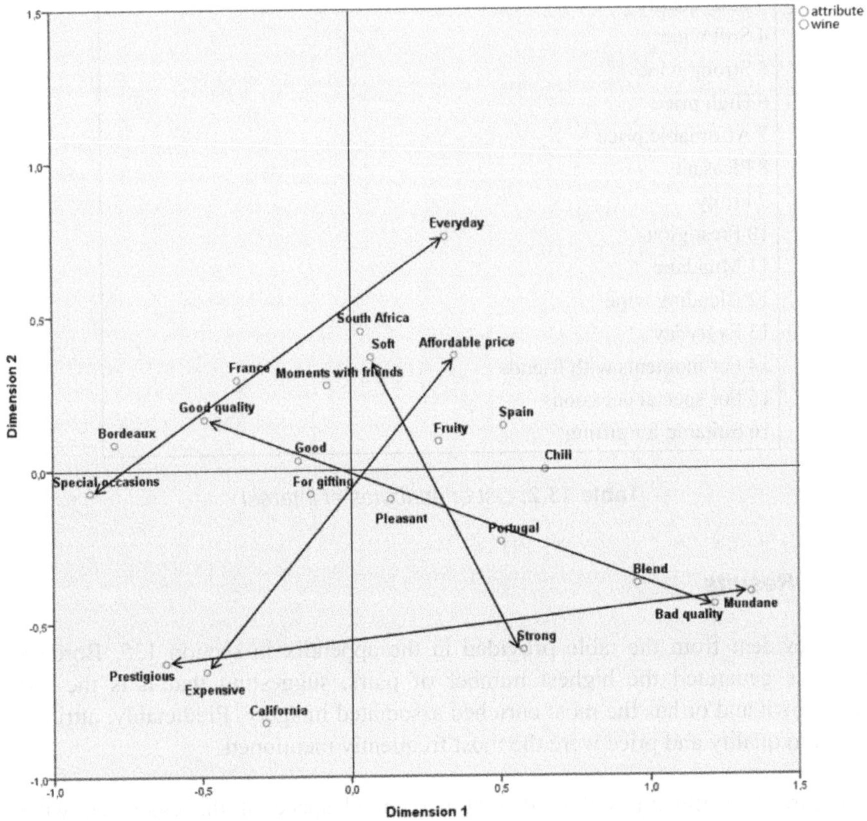

Figure 13.1. *Factorial analysis of wine-attribute matches. For a color version of this figure, see www.iste.co.uk/cheriet/wine2.zip*

In this regard, it is evident from the outset that Bordeaux wine and California wine are the farthest from the center, followed to a lesser degree by the wines from Chile, Spain and Portugal, and then France and South Africa. Similarly, the attributes furthest from the center include "prestigious", "expensive", "special occasions", "everyday", "blend", "poor quality", "mundane", "strong", "good quality", "affordable" and "soft". Few elements are near the center. No wine falls within this area. Regarding attributes, it is worth noting that "good", "suitable for gifting" and "pleasant" are the least distinguishing.

Based on this observation, we can construct an interpretation. Using the attributes, we can identify several oppositions indicated by arrows on the graph, helping us to understand the perception of each wine by the African consumer. We can observe oppositions like "prestigious/mundane", "special occasions/everyday", "good quality/poor quality", "expensive/affordable" and "soft/strong". This allows us to structure this map with a "southwest" block (prestigious, expensive, special occasions) opposed to an "affordable, mundane, everyday" block and a "northwest" block (good quality, soft) opposed to "strong" and "poor quality".

Wines from Spain, Chile and Portugal all appear to be associated with relatively regular consumption at an affordable price, which aligns with the findings of Tsapi et al. (2009). Their positions on the right side of the space place them in a perceptual space characterized by "perceived poor quality", ordinariness and blend. In terms of other attributes, they are more centrally located, indicating intermediate price and prestige. However, it is worth noting that Spain and Portugal are vertically opposed, suggesting that Portuguese wine is perceived as "stronger" than Spanish wine.

Wines from France (except Bordeaux), California, and Bordeaux are perceived as more suitable for special occasions, prestigious, of good quality and are typically more expensive. On the horizontal axis, influenced by attributes of prestige, special occasions, price and good quality, Bordeaux wine occupies the most extreme position. It is worth noting that, in contrast to Californian wine, Bordeaux's prestige is closely tied to quality and a festive character (special occasions) without the price being a direct consequence. In contrast, Californian wine's perceived prestige appears to be inseparable from a higher price. In terms of taste, Californian wine seems stronger than Bordeaux wine, which is more centrally located on the vertical axis.

13.4. Conclusion

This exploratory work provides us with a better understanding of the unique position of Bordeaux wine in the perceptual space of African consumers. Prioritizing a qualitative approach, this research effort, as well as the initial results, have appeared to be not only encouraging but also necessary in order to commence more extensive research. The goal of our ongoing research, which now emphasizes a quantitative approach, is to develop a more coherent marketing strategy for Bordeaux wine packaging and communication targeting African markets. While Bordeaux wine appears to enjoy a positive image among African consumers, the next step is to explore the conditions that will facilitate the translation of this image into purchasing decisions. This can be achieved by examining the cultural elements related to both the consumption situation and the incorporation of distinctive aspects of African culture in Bordeaux wine packaging and communication in Africa.

13.5. Appendix

Wine	Good	Good quality	Poor quality	Soft	Strong	Expensive	Affordable price	Pleasant
Bordeaux	30	38	1	14	4	15	16	13
France	24	30	1	17	6	15	16	13
Spain	19	17	10	16	8	9	22	12
Portugal	18	17	8	13	15	7	18	16
Chile	19	13	5	10	15	7	17	11
South Africa	26	28	3	16	15	9	20	13
California	23	23	4	9	16	21	9	12
Total	159	166	32	95	79	83	118	90

Wine origins	Fruity	Prestigious	Mundane	Blend	Everyday	Moments with friends	Special occasions	Suitable for gifting	Total
Bordeaux	8	16	1	1	9	16	24	12	218
France	4	7	1	5	13	13	12	11	188
Spain	10	6	9	4	11	11	5	7	176
Portugal	7	7	10	7	7	14	8	8	180
Chile	7	4	9	8	15	8	5	9	162
South Africa	9	6	4	1	16	19	7	4	196
California	6	13	4	3	4	10	10	7	174
Total	51	59	38	29	75	91	71	58	1,294

Table 13.3. *Wine/attribute association*

13.6. References

Baudouin, J. (2022). Bordeaux : les lueurs de l'export au bout du tunnel. *La Revue du Vin de France* [Online]. Available at: www.larvf.com/bordeaux-les-lueurs-de-l-export-au-bout-du-tunnel,4803197.asp.

Behaeghel, J. (1991). *Brand Packaging: The Permanent Medium.* Architecture Design and Technology Press, London.

Brent, R. (2009). A cost-benefit analysis of a condom social marketing programme in Tanzania. *Applied Economics*, 41(4), 497–509.

Brouté, N. (2016). Le marché du vin en Afrique : entre mythes et réalités. Le cas du Nigéria. In *Colloque Interdisciplinaire "Les Territoires du Vin"*, Angers.

Brouwer, R., Job, F.C., van der Kroon, B., Johnston, R. (2015). Comparing willingness to pay for improved drinking-water quality using stated preference methods in rural and urban Kenya. *Applied Health Economics and Health Policy*, 13(1), 81–94.

Caballero, R. (2009). Cutting across the senses: Imagery in winespeak and audiovisual promotion. In *Multimodal Metaphor*, Forceville, C. and Urios-Aparisi, E. (eds). De Gruyter, Berlin.

Celhay, F. and Trinquecoste, J.-F. (2015). Package graphic design: Investigating the variables that moderate consumer response to atypical designs. *Journal of Product Innovation Management*, 32(6), 1014–1032.

Chaney, I.M. (2002). Promoting wine by country. *International Journal of Wine Marketing*, 14(1), 34–40.

Charters, S. (2006). *Wine and Society: The Social and Cultural Context of a Drink*. Butterworth-Heinemann, London.

CNIV (2019). Mieux comprendre la déconsommation de vin. L'analyse approfondie [Online]. Available at: www.intervin.fr.

Dadzie, K.Q., Winston, E., Afriyie, K. (2003). The effects of normative social belief systems and customer satisfaction on rural savings programs in Ghana. *Management Decision*, 41(3), 233–240.

Darbon, D. (2012). Classe(s) moyenne(s) : une revue de la littérature. Un concept utile pour suivre les dynamiques de l'Afrique. *Afrique contemporaine*, 4(244), 33–51.

Dubihlela, J. and Ngxukumeshe, T. (2016). Eco-friendly retail product attributes, customer attributes and the repurchase intentions of South African consumers. *The International Business & Economics Research Journal*, 15(4), 163–174.

Duquesnois, F., Dikmen, L., Cheriet, F. (2021). Dynamiques concurrentielles autour de BGI (groupe COPAGEF) : acteur principal des boissons en Afrique de l'Ouest. In *La dynamique concurrentielle : acteurs singuliers, stratégies plurielles*, Bensebaa, F. (ed.). Éditions EMS, Caen.

Favier, M., Pantin-Sohier, G., Celhay, F. (2018). Is using ornaments still a crime? Package design complexity and brand perception with application to champagne labels. In *Marketing Transformation: Marketing Practice in an Ever Changing World*, Rossi, P. and Krey, N. (eds). Springer, Berlin.

FranceAgriMer (2021). Bilan 2020 du commerce extérieur des vins et spiritueux. Report, Les données de FranceAgriMer.

Gait, A. and Worthington, A.C. (2015). Attitudes of libyan retail consumers toward Islamic methods of finance. *International Journal of Islamic and Middle Eastern Finance and Management*, 8(4), 439–454.

Hall, J. and Lockshin, L. (2000). Using means-end chains for analysing occasions-not buyers. *Australasian Marketing Journal*, 8(1), 45–54.

Hirche, M. and Bruwer, J. (2014). Buying a product for an anticipated consumption situation. *International Journal of Wine Business Research*, 26(4), 295–318.

Jacquemot, P. (2012). Les classes moyennes en Afrique. Introduction thématique. *Afrique contemporaine*, 4(244), 13–15.

John, A. and Brady, M.P. (2011). Exploration of the dimensionality of the consumer ethnocentric tendencies scale in Mozambique. *Journal of African Business*, 12(1), 114–132.

Joly, M. (2005). *Introduction à l'analyse de l'image*. Armand Colin, Paris.

Kassous, R. (2019). À la conquête de l'or rouge. *Forbes Afrique*, 60, 56–59.

Lehrer, A. (2009). *Wine and Conversation*. Oxford University Press, Oxford.

Ltifi, M., Hikkerova, L., Aliouat, B., Gharbi, J. (2016). The determinants of the choice of Islamic banks in Tunisia. *The International Journal of Bank Marketing*, 34(5), 710–730.

Magnier, L. and Crié, D. (2015). Communicating packaging eco-friendliness: An exploration of consumers' perceptions of eco-designed packaging. *International Journal of Retail and Distribution Management*, 43(4/5), 350–366.

Martinez-Carrasco, L., Brugarolas, M., Martinez-Poveda, A. (2005). Quality wines and wines protected by a designation of origin: Identifying their consumption determinants. *Journal of Wine Research*, 16(3), 213–232.

Mbiti, J.S. (1990). *African Religions and Philosophy*. Heinemann, Portsmouth.

Ngoma, M. and Ntale, P.D. (2015). Perceived brand quality in uganda's pharmaceutical industry: The role of country of origin, marketing orientation and brand affordability. *Advances in Management and Applied Economics*, 5(6), 1–11.

Novixoxo, J.D., Mills, N.A., Anning, L. (2019). The effect of packaging on perceived quality and purchase intention of made-in-Ghana brands. *European Journal of Business and Management*, 11(5), 74–88.

Nugent, P. (2021). Alcohol in Africa. African Studies Centre Leiden [Online]. Available at: www.ascleiden.nl/content/webdossiers/alcohol-africa.

Ogbeide, O.A. and Ele, I. (2015). Segmenting the Nigerian wine market using combination of segmentation methods. *Mayfair Journal of Agribusiness Management*, 1(1), 76–93.

Perrouty, J.F., d'Hauteville, F., Lockshin, L. (2006). The influence of wine attributes on region of origin equity: An analysis of the moderating effect of consumer's perceived expertise. *Agribusiness*, 22(3), 323–341.

Quaye, W., Gyasi, O., Larweh, P., Johnson, P.T., Obeng-Aseidu, P. (2009). The extent of marketability and consumer preferences for traditional leafy vegetables – A case study at selected markets in Ghana. *International Journal of Consumer Studies*, 33(3), 244–249.

Saffu, K. and Walker, J. (2006). The country-of-origin effect and consumer attitudes to "buy local" campaign. *Journal of African Business*, 7(1–2), 183–199.

Thanika, D., Juwaheer, P.S., Noyaux, M.M.E. (2012). Analysing the impact of green marketing strategies on consumer purchasing patterns in Mauritius. *World Journal of Entrepreneurship, Management and Sustainable Development*, 8(1), 36–59.

Thomas, A. and Pickering, G. (2003). The importance of wine label information. *International Journal of Wine Marketing*, 15(2), 58–75.

Tsapi, V., Djeumene, P., Tchuente, M. (2009). Rôle du pays d'origine dans la perception de la qualité du vin par le consommateur africain : une étude menée dans le contexte camerounais. *Market Management*, 1(9), 38–58.

Vernette, E. (1994). La méthode de citation directe : préparer des décisions marketing efficaces. *Décisions Marketing*, 3, 101–106.

Wine Institute (2020). World wine consumption [Online]. Available at: http://wineinstitute.org/.

Yahaya, I., Yamoah, F.A., Adams, F. (2015). Consumer motivation and willingness to pay for "safer" vegetables in Ghana. *British Food Journal*, 117(3), 1043–1065.

14

Health Crisis and Wine Exports: How Should the Missions of Export Managers be Adapted?

14.1. Introduction

The global health crisis that unfolded at the beginning of 2020 has significantly disrupted international business operations: export specialists have faced travel restrictions, canceled international trade shows, challenges in prospecting and arranging overseas client visits, rising raw material costs, supply chain congestion etc. In the wine sector, this crisis context has been further complicated by the impacts of Trump's taxes and Brexit (Cheriet and Maurel 2022). As highlighted in various press[1] articles, it is crucial for export management and exporters to "reinvent themselves".

The majority of prior studies have focused on the impacts of crisis contexts at a strategic level, primarily within corporate leadership. By operating at a "professions and skills" level, this chapter presents the findings of a study carried out among a sample of French executives, managers and export sales professionals. Its aim is to understand the effects of the Covid-19 crisis on their respective occupations. This study will enable us to assess how a crisis can facilitate the reflection and adaptation of practices to evolving environmental conditions. Additionally, we shall

Chapter written by Carole MAUREL and Foued CHERIET.

1 See: https://www.usinenouvelle.com/editorial/l-export-se-reinvente-au-temps-du-covid-19.N103 1739 and https://www.lesechos.fr/pme-regions/actualite-pme/export-les-cinq-solutions-cles-pour-continuer-a-vendre-a-letranger-1279902.

identify the developed skills, commercial practices and strategic decisions that should be maintained.

14.2. Analysis of export management using a resource and competence-based approach

The resource-based view (RBV), attributed to the foundational work of Barney (1991), explains that a company's competitive advantage relies significantly on a set of resources and capabilities. As early as 1999, Beamish et al. demonstrated that companies achieving success in export operations through dedicated personnel possessed a competitive edge. These employees possess and develop "rare", non-imitable and less transferable skills according to the RBV framework. These skills encompass the ability to analyze and monitor international markets and competitors, establish logistics, adapt products for international markets, manage complex cross-cultural relationships and handle export-related risks effectively.

The RBV approach also considers "combinations" of resources as specific skills and dynamic capabilities in support of the organization's objectives (Teece et al. 1997; Winter 2003). In the context of exports, the resources and skills developed through international activities will merge with those deployed in domestic markets. This facilitates learning and replication processes in areas such as communication tools, product development, prospecting and commercial monitoring, team management and human resources. In turbulent environments marked by technological disruptions, crises or significant institutional changes, the development of these dynamic capabilities takes on particular importance (Simon et al. 2007).

Within wine companies, significant efforts have been made to develop additional skills or new capabilities to adapt their business strategies. These adaptations include adjustments in distribution channels, customer monitoring tools and communication methods. These strategic adaptations to changing contexts have been studied in various countries using different approaches. For example, studies have examined the strategies of Spanish companies in response to Brexit (Alonso et al. 2020), strategies of New Zealand firms during the financial crisis (Yu and Lindsay 2016), responses to the health crisis in Italy (Gerini et al. 2021) and the adaptation of export strategies of French wine companies to address challenges like Brexit, Trump's tariffs and the impacts of the Covid-19 crisis (Cheriet and Maurel 2022). These studies also consider how companies react based on their market orientation, whether they focus on mass markets or niche markets (Hammervoll et al. 2014), among other aspects.

Beyond companies' strategic choices, which are often constrained by institutional or logistical requirements, crises also provide an "opportunity" to test new resilience capacities within firms. This can involve the development of novel combinations of resources and skills or the acquisition of additional skills (Wang 2008; Wenzel et al. 2020). This process often includes phases of rapid learning, enhancements to quality systems and *supply chains*, as well as the recruitment or training of existing human resources (Lopez-Rodriguez et al. 2018).

One of the trends that has emerged with the recent global health crisis is the development of digitalization: remote meetings, the growth of online sales and training. Fiedler et al. (2021) explain that this digitalization is part of the development of new skills and capabilities. It has proven useful not only for maintaining communication among company employees but also for staying connected to customers and business partners in foreign markets, as well as for expanding online sales.

The development of digitization has allowed for overcoming certain constraints caused by border closures and travel limitations, as well as the discontinuation of some distribution channels (for instance, cafes, hotels, restaurants in the wine sector). In a recent study covering, among other sectors, food and beverages, Elia et al. (2021) analyzed the role of a company's resources in supporting internationalization through digital channels, especially e-commerce, by applying the RBV approach.

14.3. Crisis context and results of the French wine exports

The wine sector is one of France's top three trade surpluses, following the aerospace industry and preceding perfumes and cosmetics. Historically, France is among the world's three leading wine exporters, alongside Italy and Spain. Its primary markets, in order of significance, include European Union countries (25% of exports) and non-EU countries (75%, led by the United States, the United Kingdom, and China), as reported by the Federation of French Wine and Spirits Exporters (FEVS).

The year 2021 witnessed a recovery in the sales of French wines abroad after the slowdown in 2020, and has even surpassed pre-crisis figures. According to data from FEVS (see Figure 14.1), exports in 2021 increased by 27.5% compared to 2020 (13.1% compared to 2019) in value and by 10.6% compared to 2020 (4.2% compared to 2019) in volume. This suggests that these exporters have demonstrated resilience and adaptability in the face of an unprecedented crisis, together with other disruptions in the international environment of these businesses, including a decline in consumption in China and tensions between China and Hong Kong, Brexit for which an agreement was reached only by the end of 2020 between the European Union and the UK, although the referendum took place in June 2016, and the extra taxes imposed by the US administration on various European products, including wine, from October 2019 to March 2021 (Cheriet and Maurel 2022).

The analysis of wine exports has often focused on the determinants of performance or the management of the export process. Previous studies have also been predominantly at the macroeconomic or country level, with limited analysis of export management (Maurel and Cheriet 2020). Recently, new research has been conducted to incorporate the dimensions of crisis contexts, thereby creating a new field of research (Bundu et al. 2017), with numerous applications, albeit mostly at the macroeconomic level or within company strategies (Canavati et al. 2020). However, understanding the effects of crisis contexts on changes at a more managerial level could shed light on resource allocation, the development of new skills and the dynamics of export management capabilities.

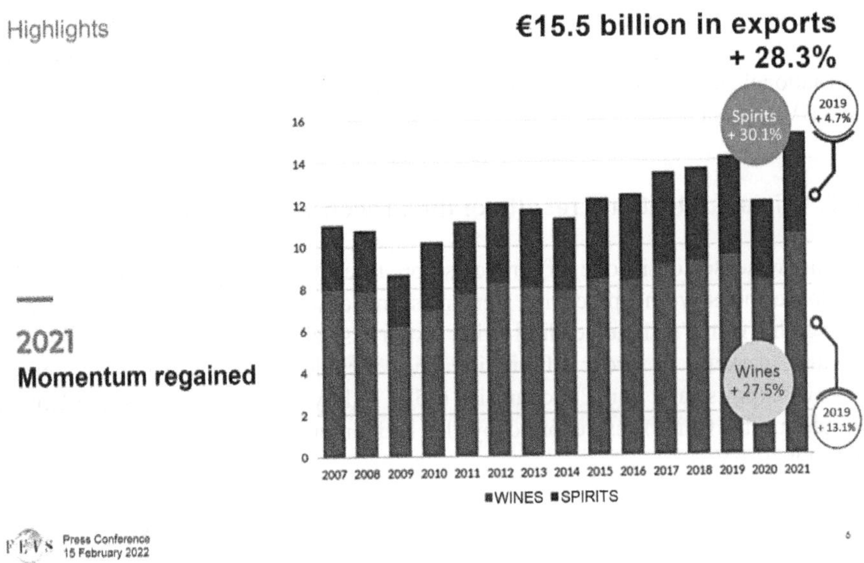

Figure 14.1. *Evolution of French wine and spirits exports from 2007 to 2021 in billions of euros (FEVS data). For a color version of this figure, see www.iste.co.uk/cheriet/wine2.zip*

14.4. Wine export management tasks

14.4.1. *A dual mission: prospecting and loyalty*

Despite the diversity of roles (export director, zone or country manager, managers in the country of origin or internationally, brand managers, product managers or distribution channel managers, export assistants and salespeople), wine export management operates at two main levels, each requiring distinct skills and conditions. The first mission involves the acquisition of new clients, primarily conducted during international trade shows attended by the sales teams. This mission includes tasks such as listing potential clients, canvassing and initiating contact. The second mission revolves around retaining existing customers, necessitating regular communication and even in-person visits.

Another approach to this profession involves the complementarity between, on the one hand, internal administrative work in collaboration with clients, and on the other hand, fieldwork with clients and prospects. This fieldwork is organized around three main phases: trip preparation (initiating contact, tour organization), actual trips (customer visits and/or trade shows) and post-trip evaluation and monitoring. These phases are essential components of the sales profession. Moreover, there is also a "brand ambassador" aspect that the field teams handle. At the management and zone or circuit management levels, team management responsibilities are added. Finally, alongside these roles, there are cross-functional tasks related to commercial tool development, regulatory compliance (standards), international risk management, process requirements and logistical constraints.

14.4.2. *Specific skills in wine export*

The unique characteristics of the wine industry, its cyclical nature centered around the vine's life and harvest and the wine production process consequently result in a cyclical pattern for commercial activities. The early part of the year is dedicated to tastings and price setting, followed by February to May, which sees the hosting of national and international trade shows. Customer retention efforts extend throughout the remaining periods, with a focus on the late post-harvest period to drive sales.

Additionally, with regard to an agricultural product, wine export management demands relatively high-level skills concerning international regulations and the presentation of both intrinsic and extrinsic technical attributes of the product. This encompasses various aspects, from the production methods to vine characteristics, oenological choices and packaging. Such complexity underscores the demanding nature of international representation missions and typically necessitates a combination of technical and commercial skills, as well as expertise in export and logistics. The importance of these specific skills in determining the success of wine export management missions varies based on the roles and regions covered (refer to Table 14.1).

<table>
<tr><th>Levels</th><th>Responsibilities</th><th>% time*</th><th>Skills</th><th>Key points</th></tr>
<tr><td rowspan="4">Export management</td><td>Team management</td><td>25</td><td rowspan="4">– Technical knowledge of the product
– Market regulation and logistics</td><td rowspan="4">– Travel time and preparation/feedback meetings
– Direct team and customer contacts</td></tr>
<tr><td>Prospecting</td><td>25</td></tr>
<tr><td>Visits/loyalty</td><td>25</td></tr>
<tr><td>Trade fairs</td><td>25</td></tr>
<tr><td rowspan="3">Manager zone</td><td>Prospecting</td><td>25</td><td rowspan="3">Area-specific knowledge (demand, competition, consumption expectations)</td><td rowspan="3">– Preparation and travel time
– Pressure from business objectives</td></tr>
<tr><td>Visits/loyalty</td><td>50</td></tr>
<tr><td>Trade fairs</td><td>25</td></tr>
<tr><td rowspan="2">Sales representative</td><td>Prospecting</td><td>50</td><td rowspan="2">– Technical knowledge of the product and logistics
– Need for renewal of the offering</td><td rowspan="2">– Pressure from business objectives
– “Internal” competition</td></tr>
<tr><td>Visits/loyalty</td><td>50</td></tr>
<tr><td rowspan="3">Export assistant</td><td>Visits/loyalty</td><td>25</td><td rowspan="3">Technical knowledge of the product, customers, and internal business tools</td><td rowspan="3">– Business follow-up
– Order management</td></tr>
<tr><td>Business follow-up</td><td>50</td></tr>
<tr><td>Administrative management</td><td>25</td></tr>
</table>

*Average estimate based on previous studies (Maurel and Cheriet 2020)

Table 14.1. *Missions and skills in wine export management*

14.4.3. *Complexity of time and resource management*

Alongside the diversity of missions and the need to adapt specific skills to the wine sector, wine export management is also characterized by the importance of time management and resource allocation, often resulting in significant trade-offs within the leadership of exporting companies. Prospecting operations can be expensive, risky and highly disruptive to the financial stability of exporting SMEs (small and medium-sized enterprises). A former export manager at Listel[2] believed

2 Interview at a conference on wine export, Montpellier Wine International Business (MWIB), Institut Agro Montpellier, February 2019.

that taking the risk of losing tens of thousands of euros was necessary to enter the American market.

Export management also entails significant time management pressures. A leader in a wine trading company exclusively focused on exports in Languedoc Roussillon[3] acknowledged that this was the primary constraint of the export manager's role in the wine industry. He explained the need to strike a balance between maintaining closeness to customers (visits, on-site tastings, participation in trade fairs) and the administrative and commercial management of these international operations (preparing and evaluating trips, achieving sales objectives, monitoring commercial performance, etc.).

14.5. Methodological approach

The diversity of missions, the specificity of skills and the development of new skills resulting from the adaptation of wine export management to the constraints of crisis contexts illustrate the complexity of our research subject. Our empirical approach follows an exploratory qualitative path. To do so, we examined the changes in the missions and skills of French wine-exporting companies. The starting point of this research, which began in September 2020, was an "alert" issued by many wine companies in the Occitanie French region, reporting the challenges of export management in a crisis context and ongoing considerations to identify the commercial and managerial practices to sustain after returning to normal. Consequently, we decided to conduct this research, with the uniqueness of commencing from the practices carried out by wine export managers and salespeople.

14.5.1. *Data collection*

Our research is built on primary data collected between March and June 2021, through semi-structured phone interviews and remote focus groups (via Zoom). This phase followed a collection of secondary data, including previous studies and monographs on wine export, scrutiny of company websites, a review of specialized press and an examination of official statistics concerning French wine export in crisis contexts (refer to Table 14.2 for a breakdown of the stages).

The primary information gathered during the interviews focused on the roles of a sales representative or an export manager in a "normal" situation, the changes brought about by the crises, including the health crisis, digitization and wine export

3 *Idem*.

management, as well as the practices implemented during the crisis that should be continued. The focus groups further complemented this data by providing accounts of real-life situations, a more precise identification of sector-specific skills in the wine industry, an exchange of experiences among participants and the emergence of discussions on practice sustainability.

Stages	Description	Objectives
Literature research	– Analysis of previous studies on the determinants of French exports of wine – Consultation of monographs and websites of French wine-exporting companies – Specialized press review and statistics of national (FranceAgriMer), international (OIV) and professional (FEVS) organizations	– Determination of generic export skills specific to the wine sector – Analysis of the effects of the crisis in terms of exports, identified by previous studies – Identification of important issues raised by companies
Interviews	– 11 semi-structured interviews lasting 1.5 hours each, with a guide around 6 themes (February–March 2021) – Separate profiles of companies (large versus SMEs, producers versus trading, independent versus cooperatives, exclusive versus marginal, different targeted export areas) and varied functions of respondents (directors, managers, sales representatives, assistants)	– Accuracy of missions/wine export management in "normal" times, evolution with crisis context, additional skills, sustainability of changes – Comparison according to company profiles and respondent functions
Focus groups	– 2 focus groups (5 and 6 representatives) in April and June 2021, with exporting companies – 2 x 2 hours of discussion around 4 main themes: wine export management, effects of the crisis, digitalization, sustainability of actions	– Confirmation of interview responses – Emergence of skills not identified during interviews – Exchange of experiences and identification of specific situations – Identification of specific skills and the effects of the crisis context

Table 14.2. *Methodological stages and data sources*

14.5.2. *Respondent sample description*

We intentionally created a diverse group of participants, excluding some duplicated profiles. This sample was designed to provide the most comprehensive representation of the observed situations. Table 14.3 displays the sample used in our study and showcases the sizes of the relevant companies, respondent profiles, their country focus and export assignments.

Corporate names	Business	2020 turnover*	% export turnover	Respondent's role	Country focus
Château Campuget	Production	3–4	70	Export zone manager	Germany, Holland, France
Champagne Baron Fuentes	Production	15	50	Export manager	Large-scale export
LGI Wines	Trade	75	100	Export manager	EU, US, Asia
Advini (1)	– Production – Cooperative – Trade	248	50	On trade zone manager	Benelux, Denmark
Sieur d'Arques	– Production – Cooperative	40	45	– Export manager – UK commercial	EU, US, Asia, UK
Confrérie des Domaines	Trade	11	100	Export sales assistant	UK, Ireland, Africa, India
Advini (2)	– Production – Trade	248	50	– Traditional circuit manager – Sales representative	UK
Murphy Wine Company	Brokerage	2	80	Associate director	US and other countries
Bonfils	– Production – Trade	20	60	– Export sales representatives – Zone manager	Germany and Switzerland
Puech Haut	Production	20	30	Export manager	US, Asia
Alexander Krossa	Trade	4.5	100	Export sales representatives	EU

*Turnover (in millions of euros)

Table 14.3. *Sample and respondent description*

The 11 interviewees, who also took part in two separate focus group sessions (called FG1 and FG2), occupy various roles in wine export management (export director, sales, zone manager, export assistant, etc.), and cover different geographical areas (large-scale export versus Europe, specific countries or zones, etc.). These interviewees also have significant variations in terms of their previous international experiences, ranging from several months to approximately 20 years. Their respective parent companies also exhibit a broad diversity in terms of legal structures (cooperatives, independent winegrowers), sizes (annual turnover ranging from 3 to nearly 250 million euros) and activities (production, trade, and brokerage). The surveyed companies have distinct international profiles, with export levels varying from 30% to 100% of their total revenue and different levels of dedicated export staff. Lastly, these companies originate from various wine production regions, all located in France: Languedoc-Roussillon, Côtes du Rhône, and Champagne. Finally, the most prominently represented export regions, in order of significance, are EU countries, the United States, followed by the United Kingdom and Asia, which are equally significant.

14.6. Results and effects of the crisis on wine export management

Before delving into the results regarding changes implemented in a crisis context and the sustainability of certain activities, our interviews broadly confirmed the previously described distribution of working time and the complexity of the export manager's role:

> Every fourth week involves traveling, regular hosting of clients on the properties, ongoing prospecting via email, telephone, and in the field, regular promotional activities, offer monitoring, participation in trade shows, strategic organization within the designated area, management of the representation approach in the region, direct or indirect trade depending on the country, oversight and management of sales agents, payment monitoring, goal setting, and order tracking. [...] It's an enjoyable profession that allows for travel and engagement with diverse cultures, facilitates language practice, provides opportunities for meaningful encounters, and requires a results-oriented mindset, suitable for those who enjoy challenges. However, it is also a demanding profession: preparation and organization of trips, time spent in transit, intervals between appointments, extensive travel, environmental concerns, time zone differences, fatigue, and stress. This profession demands good physical condition and personal sacrifices (being away from family) (Respondent D, FG1).

14.6.1. *Changes in export management*

Overall, our interviews and the two focus groups have helped identify various types of changes. Firstly, these changes pertain to our perception of time. Secondly, they involve shifts in customer and market relationships. Lastly, they concern the digitalization of processes. These three dimensions encompass the development of new skills, as well as the amalgamation of skills necessitated by the crisis context with the preexisting capabilities. They also underscore the significant adaptability and resilience of both teams and leaders through a rapid learning process. Consequently, these skills are intrinsic to the role of the export manager:

> The main characteristic of our role as export sales representatives is to be responsive and adaptable. The new context we encountered in 2020 compelled us, even more so, to demonstrate agility and responsiveness in facing the challenges of this situation: rapidly learning new communication tools (Teams, Zoom, Skype, etc.) (Respondent K, interview).

However, it is important not to ignore certain business challenges resulting from health and market constraints. The survey participants reported losses of customers and market share, as well as withdrawals from countries or distribution channels.

> We had to adjust to the closure of certain channels (such as HORECA in France) to re-engage with contacts in other channels. This proved to be challenging in several European markets, particularly with pressure on targets (a reduction of –50% in certain destinations) (Respondent F, FG2).

In certain instances, these effects have led to added pressures on specific channels, logistics and supply bottlenecks, and clear challenges in gathering information and obtaining customer feedback, which were not alleviated by the utilization of digital tools:

> The most significant impacts of the Covid-19 crisis were an increase in demand from supermarkets and pressure on raw materials, logistics, and transportation providers. Two other effects have been noted: a substantial reduction in in-person business meetings, but this was offset by more digital contacts (videoconferencing) or telephone communications to sustain relationships with customers and partners. The ability to engage in and gather business information during remote interactions was more challenging. This was only partially replaced by digital means (Respondent H, FG2).

Many companies also mentioned seizing numerous opportunities in the context of the crisis, especially by diversifying customers and markets and adapting their products:

> The impact of Covid is real, but there is a significant shift in sales towards the internet and supermarkets. Wine consumption has either remained stable or decreased slightly, depending on the country. I recall that, over the past year, salespeople who persevered managed to maintain their sales. Being innovative is sometimes necessary, and having a strong rapport with customers is likely crucial. It's also essential to adapt to new communication tools like Zoom, social networks, and video sharing (Respondent G, FG2).

The second change pertained to the management of time and the optimization of travel. This was a shared characteristic among all of the respondents, evident in both the interviews and focus groups. Among the numerous accounts, here is one from an export manager (Asia) that illustrates the achieved benefits:

> The new context has enabled me to better manage my time. I've learned to market effectively while minimizing time-consuming travels and the associated costs. With reduced travel, sales methods have adapted, and in some cases, improved. Another noteworthy observation is that extensive travel is not always necessary for successful sales. A portion of the business activities can be conducted remotely. In this regard, the forced development of these new methods has been advantageous for our company's management (Respondent D, FG1).

Lastly, the third significant change involved the digitization of certain wine export management tasks:

> The crisis has allowed the wine industry to rejuvenate in the e-commerce and digitalization sectors. It has altered some perspectives, potentially impacting younger consumers. While hypermarkets (GD) have greatly benefited from this crisis, others have still managed to seize opportunities during these challenging times (Respondent B, FG1).

Beyond the development of new tools, it is the integration of digitalization into the wine industry, with all future growth possibilities (communication, e-commerce, telecommuting, etc.), that is emphasized.

This was confirmed by numerous respondents, with the added consideration of the future challenges of wine export, such as the widespread adoption of certification and traceability:

> The Covid pandemic compels us to alter our routines, with videoconferencing taking precedence over air travel. We've opted to visit in person once a year and handle the rest through video. Samples will be sent on our behalf, as is currently the case. Certification gains significance as human interactions become less frequent, and quality procedures will play a more prominent role in enhancing quality and traceability (Respondent H, interview).

Table 14.4 provides an overview of certain results observed concerning the organization of missions, the specific skills developed in a crisis context and the associated key points.

14.6.2. *Advantages and limitations of the new context*

The second point highlighted by the interviews and focus groups revolves around the perceived advantages and disadvantages for wine export professionals in the context of the crisis. The advantages encompass cost and time savings, as well as the utilization of digital tools. There is also a fresh perspective on the two primary missions (prospecting and customer loyalty), the opportunities presented by new markets and distribution channels and the reinforcement of social connections.

In addition to the savings made due to the inability to travel internationally during the lockdown, another cost reduction was related to the absence of trade shows. This has led to the development of new event-related skills within the teams:

> We had to discover alternative methods of meeting. Furthermore, the pressure on specific channels and markets compelled companies to explore different avenues, especially the internet and other export destinations, to make up for it (Respondent C, FG1).

The crisis context has enabled the identification of specific market opportunities and the development of more efficient and innovative digital tools. Time savings have also contributed to enhancing the teams' collaborative work skills:

> The time that was not used for travel was effectively utilized for human resource management. In our company, we faced a significant challenge and engaged in extensive collaboration with our social partners, which strengthened social connections. This applied not only

to the cooperators and their employees but also involved imparting vision to employees who were lacking it. Many companies will emerge from this experience with stronger social bonds (Respondent K, FG 2).

<table>
<tr><th>Levels</th><th>Task</th><th>% time*</th><th>Additional credentials</th><th>Key points</th></tr>
<tr><td rowspan="4">Export management</td><td>Management of teams</td><td>25</td><td rowspan="4">– Management of teams working remotely
– Digital and communication
– Digitized business tools</td><td rowspan="4">– Gains in travel time
– Digital tools, prospecting innovation, virtual fairs
– Importance of maintaining direct team and customer contacts</td></tr>
<tr><td>E-prospecting</td><td>25</td></tr>
<tr><td>E-visits</td><td>25</td></tr>
<tr><td>Virtual tradeshows</td><td>25</td></tr>
<tr><td rowspan="3">Area manager</td><td>E-prospecting</td><td>25</td><td rowspan="3">– Digital and communication
– Digitized business tools
– Collection and analysis of secondary data</td><td rowspan="3">– Gains in preparation and travel time
– Pressure from business objectives
– Adaptation of data collection tools (markets, competitors) to the absence of travel</td></tr>
<tr><td>E-visits</td><td>50</td></tr>
<tr><td>Virtual tradeshows</td><td>25</td></tr>
<tr><td rowspan="2">Sales representative</td><td>E-prospecting</td><td>50</td><td rowspan="2">Proficiency in digital prospecting and loyalty tools</td><td rowspan="2">– Remote work and teleworking and pressure
– Proficiency in digital business tools
– Adaptation of commercial tools to competitors and virtual fairs</td></tr>
<tr><td>E-visits</td><td>50</td></tr>
<tr><td rowspan="3">Export assistant</td><td>Digitization</td><td>25</td><td rowspan="3">Mastery of digital communication, commercial monitoring and administrative management tools</td><td rowspan="3">– Remote work, digital sales and order tracking tools
– Development of digital business tools (video capsules, digital brochures, etc.)</td></tr>
<tr><td>Business follow-up</td><td>50</td></tr>
<tr><td>Communication</td><td>25</td></tr>
</table>

*Average estimate based on interviews and focus groups

Table 14.4. *Change of tasks and new skills of wine export management*

Lastly, and as emphasized repeatedly by our respondents, digitalization has been seen as a fresh strategic opportunity and a required enhancement for industry stakeholders:

> The new tools were indispensable to remain active during the lockdown. We soon recognized that the wine sector was lagging behind other industries in terms of digitization. Wine is still a product that people enjoy tasting, taking their time to select, so it's challenging to replicate a similar experience online. Nevertheless, online sales websites are achieving impressive results, and consumers are discovering both familiar wines and new suggestions (Respondent I, FG2).

The drawbacks, on the other hand, pertain to social distancing, the adjustment of digital tools, and challenges associated with the particularities of the profession: product tasting, data collection in the field, interactions with clients, etc.

Respondents clearly identified the absence of connections with customers and suppliers as a source of commercial inefficiency, affecting both prospecting and loyalty tasks, as well as activities specific to the wine sector such as tastings and vineyard visits:

> We've adapted, as have our customers, but over the months, it's clear that customers long for human interaction in person. While communicating through a screen is undoubtedly efficient, something is missing in the end. And not to mention a crucial aspect of our profession, tasting – doing it via video conference remains a rather frustrating experience (Respondent K, interview).

The non-utilization of interpersonal skills was perceived differently depending on the levels of experience and export commitment of the interviewed professionals. In some cases, this social distancing has led to team discomfort, exacerbated by the pressure on business objectives, the uncertainty of the crisis context and the absence of prospects. Fewer trips and trade shows mean more free time with the family, but this absence from the field also comes with constraints: a reduced competitive intelligence on the ground and fewer opportunities to meet buyers at trade shows or during tours:

> This change of profession is not always favorable for experienced salespeople who are used to traveling. A certain gloom, or even in some cases, a real depression, has slowly set in among some of them (Respondent E, FG1).

Finally, the widespread adoption of digital tools in wine export management was not universally appreciated. It also underscores the significance of skills gained through experience and those related to market knowledge:

> The widespread use of an all-digital approach may come with several criticisms: a dilution in the quality of interactions with customers (sample tracking, non-targeted clientele); insufficient customer feedback regarding market reactions and trends; no immediate feedback from customers on products, brands, or promotional ideas, and the inability for export sales representatives to tailor their pitch in a direct conversation to address customer objections; minimal feedback from specific professionals in virtual trade shows we participated in, while receiving more responses from sommeliers and wine merchants… This could be due to issues with targeting virtual trade shows, obstacles to innovation or product launches (lack of in-depth market knowledge), and hindrances in providing customized support for customers (Respondent B, FG1).

14.6.3. *Skills to be retained after the crisis*

The third section of our findings revolves around the sustainability of certain skills developed in the context of a crisis. Respondents took a step back to identify what should be "retained". The responses center on four main areas: the persistence and enhancement of specific digital tools, the reinforcement of team management flexibility and skills, the development of cross-functional strategic thinking on crisis management and the creation of strategic intelligence tools (including climate change effects, commercial and competitive intelligence and regulatory intelligence).

Regarding the sustainability of digital tools, companies are now making trade-offs in costs and efficiency of activities:

> With hindsight, we can see that there are certain meetings that can be conducted on Zoom and it works. Everyone is eager to meet in person again, but digitization will be retained. As for trade shows, there is a genuine need. Trade shows serve a triple purpose: conducting business, building relationships, and competitive intelligence (Respondent K, interview).

Other companies are considering developing these tools using external resources:

> We hired two additional individuals with digital marketing expertise to support the sales and export teams. In other cases, we utilized service providers. For instance, we worked with Hop Wine for sending samples, arranging e-tastings, or organizing virtual shows (Respondent A, FG2).

> Despite the intention to develop or sustain certain digital tools, many respondents discussed a reorganization of wine export management, recognizing the importance of relationships in certain roles. This reorganization involves changes induced by digitization and the company's strategy. It includes plans to keep one out of every five trips per customer, eliminate non-essential in-person meetings, maintain physical travel for handling logistical or payment issues, and rethink the relevance of participating in trade shows. This might involve reducing the physical presence of sales teams in specific markets and alternating physical attendance (every other year) at certain trade shows (Respondent E, FG1).

The consideration also extends to market diversification and the pursuit of commercial proximity, which is seen by exporting companies as a means to reduce risks and manage logistical constraints:

> In terms of commercial outlets, the crisis has prompted a consideration of expanding into nearby markets: France and Southern Europe to avoid potential logistical and commercial constraints (Respondent E, FG1).

Finally, in terms of strategic thinking and the development of cross-functional team skills, it involves implementing monitoring systems as well as optimal and flexible human resource management. In this regard, the export director of a company in our sample stated that:

> Beyond the direct effects of the crisis, we have initiated three additional strategic considerations within our company: the pace of sales recovery with the reopening of CHR [Cafés, Hotels, Restaurants] and the return of tourism, as well as the pressure on wine (our trading company's primary raw material), especially after the frost episodes and related service providers. We are internally reflecting on crisis management tools, including HR flexibility, strategic market-product positioning, and adaptation capabilities. We are also contemplating tools for maintaining and enhancing customer relationships (Respondent I, FG1).

Another mentioned the establishment of a unit involving many employees of the export teams and other executives of the company to lead:

> reflections in terms of anticipating future crises and the impacts of climate change, with a focus on monitoring developments in agronomic research and emerging market trends, including resistant grape varieties, blending, planting density, etc. (Respondent E, interview).

14.7. Summary of results and discussion

The health crisis has had several effects, some beneficial and others more problematic. It has compelled teams to adapt, reinvent and innovate in their export management practices to sustain or even expand international sales. Theoretically, this corresponds to the development of new skills and their integration with the company's existing resources, leading to the creation of dynamic new skills (Simon et al. 2007).

In general, companies have shown a high level of resilience and adaptability (Alonso et al. 2020), primarily because of their teams' ability to harness their skills. They learned and replicated specific practices (Lopez-Rodriguez et al. 2018) to make new resource-, market- and product-related decisions. Some adaptations were related to export management, while others were more specific to the wine sector (tastings, samples, assemblage decisions with customers, wine merchant circuits, etc.). These findings align with those obtained by Canavati et al. (2020) and Gerini et al. (2021) in the Spanish and Italian contexts.

The beneficial effects on the export sales profession are related to the effects of the decline, or even stops in some cases, of travel abroad, and are of several kinds.

First, this relates to a different utilization of time: the introduction of remote training for sales representatives, strengthening social connections, the establishment of new distribution channels, particularly in export and e-commerce, more time allocated to maintaining relationships with the existing customer portfolio and the allocation of teams' time to broader company projects (strategic planning, implementation of a new information system, corporate social responsibility considerations, establishment of a monitoring unit, etc.). This allows for enhanced collective cohesion within the export teams and the development of "vertical" skills. It has also led to stronger ties with other local and national teams through more frequent interactions and collaborative discussions on resource allocation. Ultimately, this represents the cultivation of new cross-functional skills that extend beyond export management.

Next, there is a consideration for streamlining these travels due to their impacts on costs, time, stress and quality of life. Even though some lament their inability to travel, they also appreciate this lifestyle, which offers more family time and less stress and fatigue associated with transportation and time zone differences. Lastly, there are financial implications tied to the reduction in travel expenses. This has either resulted in cost savings or the reallocation of these expenses to other budget items or projects.

On the other hand, we have observed a series of issues arising from the reduction in travel to foreign markets: (1) the lack of on-site contact, resulting in a limited understanding of market trends and conditions; (2) the mixed reception of digitalization in certain practices, such as trade fairs and tastings, which require additional preparation time (e.g. sample shipments); (3) challenges related to decreased sales outlets, the closure of specific channels or markets and pressures on raw materials. The strategic considerations have centered on the importance of diversifying market channels, crisis anticipation and strategic monitoring skills (Bundy et al. 2017).

In all cases, it is noted that the situation experienced by these export professionals served as a genuine "stress test" or resilience benchmark for the sales teams. Export sales functions have undergone partial digitalization, necessitating the hiring of specialized recruits or service providers, as well as the mobilization of internal skills: remote meetings (both within the company and with international business partners), virtual trade shows, remote tastings, sales training, digitalization of communication materials, video capsules, and the creation or revamping of websites.

Among all of these practices tested during this health crisis, some have been retained (meetings, training, video capsules, digital communication, etc.) to adapt the profession to its new reality. However, there is a clear and unanimous desire for international travel: customer contacts, competitive and commercial monitoring. Despite the interest in certain tools, an all-digital approach is excluded from the future vision of the export manager's role. They believe that the social aspect at the core of the sales function cannot be replaced by virtual contact. Several individuals have shared their experiences of shortcomings in virtual trade shows or remote tasting sessions. Therefore, in-person trade shows, which are more focused and less numerous, as well as in-person professional contacts, are preferred.

The resource allocation, the sustainability of newly developed skills, the establishment of collective and cross-functional dynamic capabilities and skills of export management teams and the development of skills for strategic and commercial monitoring and anticipation of future crisis episodes (climate change, logistics and raw materials) are fully aligned with the RBV approach and the field of international crisis management (Wang 2008; Wenzel et al. 2020). This underscores the relevance of combining these two research streams in future investigations.

Levels	Changes made	Sustainability
Relation to time	– Gains in time (travel, preparation, pre- and post-travel meetings) – Analysis of work rhythm pressure	– Some remote tasks and maintaining others in physics – Reflections on the organization of the work of export managers
Relation to tasks	– Analysis comparative analysis of the efficiency of travel and participation in trade fairs – Adaptations and importance of customer relations	Maintenance of a minimum threshold of field visits and customer contacts
Additional credentials	Digital, communication, remote loyalty	Strengthening the skills of teams
Financing	– Savings achieved: travel, prospecting, costs of participation in trade fairs – Arbitration of the commercial actions to be carried out	– Reallocation of certain resources – Procedures for comparing the effectiveness of commercial tools
Human resources management	– Adaptation of teams – Job flexibility – Pressures on objectives	– Recruitment on cross-functional functions (digital, crisis management, etc.) – Team training (remote work, digital)

Table 14.5. *Crisis context and skills of wine export management*

14.8. Conclusion

Frequently examined from a macroeconomic standpoint or in the context of corporate international strategies, export management has been approached in this chapter through the lens of resource and competence-based approaches. It has also been analyzed in the context of international crisis management, considering the new skills developed by export professionals during the 2020 health crisis. Consequently, we have explored the teams' adaptability and resilience, as well as their ability to leverage existing skills to cultivate new dynamic capabilities and competences.

Our empirical research centered on French wine companies actively involved in export activities. In this study, we have employed an exploratory qualitative approach, collecting primary data through 11 interviews and 2 focus groups. This primary data was complemented by secondary data obtained from national and international wine institutions. Our investigation has delved into the realm of wine export management professions, identifying the core functions and missions during "normal" times, as well as their evolution within the context of a health crisis. Ultimately, we interviewed these professionals to uncover the competences and practices that should be maintained and to draw key insights regarding wine export management.

Our findings revealed several effects of the crisis context: changes in the perception of time, a reassessment of relationships with customers and markets and a movement towards the digitization of certain tools and functions. Moreover, we observed strong resilience and significant adaptability, which prompted companies to make resource-related decisions and develop new skills either internally or by engaging external partners and providers. Many of these skills are not exclusive to the wine industry, meaning that these results could resonate with situations in other sectors. Finally, this context drove many companies to develop both vertical and cross-functional skills in wine export management and other commercial functions.

Our research is not without limitations, both theoretical and empirical. The incorporation of alternative approaches could strengthen the relevance of the findings. For instance, introducing entrepreneurial perspectives (export orientation) or marketing strategies (strategy and deployment) could complement these preliminary results. Likewise, broadening the sample size or conducting more extensive surveys should lead to substantial confirmatory analyses. Applying these findings to other sectors or countries would enable result comparisons and the identification of sector-specific skills.

Despite these limitations, this work carries various managerial implications. Firstly, it emphasizes the need for companies to harness their existing skills and combine them to develop new ones, with the aim of replicating these competencies in other functions. Secondly, it underscores the importance of flexible yet specialized profiles for wine export professionals, including technical product knowledge, adaptability and on-the-ground connections. Thirdly, it highlights the significance of maintaining social connections, both with customers and among export management teams, through regular remote interactions and training programs that facilitate learning and skills development. Lastly, the managerial implication involves the establishment of reflection and monitoring units to anticipate future crises (related to raw materials, climate change, logistics) and to manage teams more cohesively. This proactive approach is essential for the optimal management of wine exports.

14.9. References

Alonso, A.D., Seng, K., Seamus, O. (2020). Brexit, the wine sector and organizational resilience: An exploratory study of Spanish Wineries. *Review of International Business and Strategy*, 30(3), 301–322.

Barney, J. (1991). Firm resources and sustained competitive advantage. *Journal of Management*, 17(1), 99–120.

Beamish, P.W., Karavis, L., Goerzen, A., Lane, C. (1999). The relationship between organizational structure and export performance. *MIR: Management International Review*, 37–54.

Bundy, J., Pfarrer, M., Short, C.E., Coombs, W.T. (2017). Crises and crisis management: Integration, interpretation, and research development. *Journal of Management*, 43(6), 1661–1692.

Canavati, S., Bauman, M., Wilson, D. (2020). The wine industry & the COVID-19 pandemic. *Wine Business Journal*, 4(2), 1–4.

Cheriet, F. and Maurel, C. (2022). Les effets du contexte de crise sur les exportations de vin : étude exploratoire auprès des exportateurs français. *Management international*, 27, 75–88.

Elia, S., Giuffrida, M., Mariani, M.M., Bresciani, S. (2021). Resources and digital export: An RBV perspective on the role of digital technologies and capabilities in cross-border e-commerce. *Journal of Business Research*, 132, 158–169.

Fiedler, A., Fath, B., Sinkovics, N., Sinkovics R.R. (2021). Exporting from a remote, open economy during Covid-19: Challenges and opportunities for SMEs. *Management Perspectives on the Covid-19 Crisis*, 71–82.

Gerini, F., Dominici, A., Casini, L. (2021). The effects of the Covid-19 pandemic on the mass market retailing of wine in Italy. *Foods (Review)*, 10(2764). doi: 10.3390/foods10112674.

Hammervoll, T., Mora, P., Toften, K. (2014). The financial crisis and the wine industry: The performance of niche firms versus mass-market firms. *Wine Economics and Policy*, 3, 108–114.

López-Rodríguez, J., Calvo Dopico, D., Del Castillo Puente, A.M. (2018). Export performance in Spanish wineries: The role of human capital and quality management system. *European Journal of International Management*, 12(3), 311–333.

Maurel, C. and Cheriet, F. (2020). 10 years of academic research on wine exports is a must! A bibliographical review of publications. In *Wine Management and Marketing: Opportunities for Companies and Challenges for the Industry*, Cheriet, F., Maurel, C., Amadieu, P., Hannin, H. (eds). ISTE Ltd, London, and John Wiley & Sons, New York.

Simon, D.G., Hitt, M.A., Ireland, D. (2007). Managing firm resources in dynamic environments to create value: Looking inside the black box. *Academy of Management Review*, 32, 273–292.

Teece, D.J., Pisano, G., Shuen, A. (1997). Dynamic capabilities and strategic management. *Strategic Management Journal*, 18, 504–534.

Wang, J. (2008). Developing organizational learning capacity in crisis management. *Advances in Developing Human Resources*, 10(3), 425–445.

Wenzel, M., Stanske, S., Lieberman, M.B. (2020). Strategic responses to crisis. *Strategic Management Journal*, 42(2), 16–27.

Winter, S.G. (2003). Understanding dynamic capabilities. *Strategic Management Journal*, 24, 991–995.

Yu, Y. and Lindsay, V. (2016). Export commitment and the global financial crisis: Perspectives from the New Zealand wine industry. *Journal of Small Business Management*, 54(2), 771–797.

15

Establishing Trust in a B2B Wine Export Relationship in Africa

15.1. Introduction

Internationalization can be a survival or growth strategy for some companies and corporations, depending on the products they sell. However, entering an international market requires using strategies based on a set of key factors related to the company's specific characteristics, the institutional environment and the particularities of the target market (Meyer et al. 2009). Furthermore, the African continent is known for its complex geography, lack of infrastructure and potential economic and political instability, making it a challenging market to penetrate. Nevertheless, Leke et al. (2018) suggest that some companies located in Africa are highly profitable, and certain markets in Africa are thriving. For example, the wine market in Africa has been experiencing significant growth in recent years and is primarily an import-driven market. In this context, trust plays a crucial role in wine export sales for several reasons: (1) trust is fundamental in inter-business relationships; (2) importers must trust their suppliers concerning the appeal of their wine packaging for the African market, and these importers face the risks of establishing a local wine market; (3) wine is a unique product, and its appreciation requires skills that may not be familiar to Africans, which underscores the need to trust suppliers regarding product quality. Based on these considerations, it is crucial for wine importers in Africa to invest in finding the right business partner and establishing a long-term trusting relationship. This chapter focuses on building trust between African wine importers and their foreign business partners. It aims to answer the question of how to establish trust between wine suppliers and importers in an African context. Given that inter-business trust is influenced by the supplier's

---

Chapter written by Mathieu NLEND-MOMNOUGUI, Franck DUQUESNOIS and Jean-Philippe GALAN.

behavior towards the customer and the value proposition offered (Doney et al. 2007), our main research question can be broken down into four subsidiary questions:

– What are the effects of the factors related to the supplier's behavior in building trust?

– What are the effects of supplier supply-side factors in building trust?

– Is building trust between the foreign supplier and the African wine importer an antecedent to engaging in a long-term relationship?

– How do African cultures affect the effects of these factors leading to trust between the foreign supplier and the African importer of wine?

15.2. Trust in business relationships

Inter-organizational trust refers to a collective orientation of trust among individuals from partner firms (Zaheer et al. 1998). This is how research on trust in business-to-business (B2B) settings was established, focusing on the study of the antecedents and consequences of this trust. Consequently, three levels of analysis are observed concerning the antecedents of inter-business trust: partners, the dyad and exchanges (Donada and Nogatchewsky 2007). In the case of partners, studies address the inherent characteristics of companies, such as size, expertise and performance within their specific environment. The dyad pertains to the relationship between two companies and the factors related to shared values, interdependence in the partnership, specific investments and alignment of goals. The third level, focusing on exchanges, examines the interactions between individuals responsible for intercompany transactions. Here, antecedents of trust are associated with the course of cooperation, the quality of collaboration, opportunism, conflicts, the frequency of communication and exchanges, and the personal connections between those involved in the transactions. In our study, we analyze the development of initial trust from an African importer towards a wine supplier based in Europe. However, specific antecedents, such as conflict or opportunism, are not examined, as we are within the context of initial trust. Considering that trust in buyer–supplier relationships is also defined as "perceived credibility and benevolence" (Doney and Cannon 1997, p. 36), the context of wine importation in Africa aligns with this definition. Local importers seek to reduce perceived risk in their purchases by selecting companies they can trust from the outset: those likely to work reliably and have expressed an interest in the well-being of African businesses. Table 15.1 presents the results of research on the antecedents involved in building trust between two companies.

Antecedents of trust	
Level 1. Partner characteristics	
Performance	Wiertz et al. (2004)
Reputation	Lee and Dawes (2005)
Expertise	Lee and Dawes (2005)
Size	Doney and Cannon (1997)
Commitment of the partner	Gao et al. (2005)
Level 2. Dyad characteristics	
Interdependence	Kumar et al. (1995)
Satisfaction with past interactions	Lui et al. (2006)
Inter-business resemblance	Lui et al. (2006)
Sharing values	MacMillan et al. (2005)
Goal convergence	Anderson and Weitz (1989)
Social and structural ties	Gounaris (2005)
Level 3. Trading characteristics	
Satisfaction with the exchange	Chu and Fang (2006)
Relationship exchange	Wiertz et al. (2004)
Communication	Chu and Fang (2006)
Information sharing	Doney and Cannon (1997)
Support	Dyer and Chu (2000)
Sympathy	Doney and Cannon (1997)

Table 15.1. *Results of research on the antecedents of trust*

15.3. Antecedents of trust in buyer–supplier relationships

15.3.1. *Social antecedents of trust*

Social behaviors enabling the establishment of trust occur at level 3 of trust antecedents, which involves relational exchanges between representatives of each company. Indeed, social interactions between the parties' representatives and the maintenance of open communication are essential factors for building trust between two companies (Doney et al. 2007).

15.3.1.1. *Social interaction*

The level of interaction between a supplier and a buyer is strongly associated with the establishment of social links. When this social interaction is frequent and provides buyers with information that piques their interest in the relationship, it leads to trust (Doney and Cannon 1997). Furthermore, interaction can also enhance initial trust if the degree of perceived sympathy helps buyers ascribe benevolent intentions to suppliers (Doney et al. 2007). This is especially relevant as, with the advent of new technologies, buyers are already well informed about suppliers before making contact (Ahearne et al. 2019). New technologies facilitate increased interaction between buyers and sellers. In this regard, African importers communicate extensively with their wine suppliers through WhatsApp, which has become their most common means of communication. We thus propose the H1 hypothesis.

H1.– An African importer's trust in a wine supplier is positively linked to frequent social interaction between them.

15.3.1.2. *Open communication*

Open communications refer to formal or informal communications that take place sincerely and substantially between buyers and suppliers (Doney et al. 2007). They contribute to relationship marketing and the construction of trust (Smith 1998). Similarly, information asymmetry, where sellers conceal information about their products from buyers, is becoming less relevant in B2B, as buyers now have easy access to useful information about sellers and their products or services (Atef et al. 2020). This is evident with African importers who are increasingly researching the prices of the wines they seek before even contacting a specific supplier. We thus propose the H2 hypothesis.

H2.– An African importer's trust in a wine supplier is positively related to the degree of open communication with the supplier.

15.3.2. *Antecedents of trust related to the supplier's offer*

Offer-related trust history refers to (1) the overall quality of service from the supplier and (2) the value perceived by buyers.

15.3.2.1. *Overall quality of service*

Quality of service is defined in various ways in the literature. Among these, Parasuraman et al. (1985) defined it as "the provision of excellent service or service that exceeds customer expectations". Given that we are in a B2B context, this definition appears to be well suited to the study's context. Furthermore, since wine import involves both the expected quality of the purchased product (wine) and the service that accompanies its supply, the conceptualization by Parasuraman et al. (1988) was used to evaluate service quality in this study, comprising five dimensions: reliability, helpfulness, assurance, empathy and tangibility. Therefore, the aim is to understand how African importers assess the quality of service from their wine suppliers based on these five dimensions.

15.3.2.1.1. Reliability

In this context, reliability can be defined as the supplier's company's ability to fulfill its commitments regarding delivery, service provision, conflict resolution and more, to their customers. It is a crucial element in the perceived quality of service by the customer. For African wine importers, reliability includes meeting delivery deadlines, ensuring the availability of sought-after wines and effectively addressing any issues that may arise.

15.3.2.1.2. Helpfulness

This dimension refers to the attitude and responsiveness of the supplier in addressing customer requests, concerns and complaints. In the literature, numerous studies assess a company's helpfulness through its personnel. Therefore, it is essential for customers to sense the engagement of the supplier's employees when making requests. This aspect becomes even more significant in the African context, as it is a continent characterized by collectivism and feminism. Warm human interactions, flexibility and conflict resolution through negotiation are prevailing social behaviors (Hofstede 2001).

15.3.2.1.3. Assurance

This dimension defines the supplier company's and its employees' ability to inspire confidence in customers. This dimension is crucial in the context of wine importation into Africa for at least two reasons. The first reason is related to the payment methods offered by customers in Africa. It is quite common for some importers to prefer paying for their orders in installments due to their lack of complete trust in the supplier, making them hesitant to pay the full amount upfront. The second reason is that a significant portion of African wine importers lacks extensive knowledge of wine and, therefore, must rely on their suppliers' trustworthiness regarding the quality of the wines they purchase. This dimension heavily emphasizes the knowledge and skills of employees and the security provided by the company.

15.3.2.1.4. Empathy

Empathy in the evaluation of service quality refers to the individualized attention that customers receive from the supplier's company. This entails the positive impact of personalized services that make customers feel unique and special to the company. Special attentions, like offering a few bottles of wine or providing discounts on the price, are typically highly appreciated by Africans.

15.3.2.1.5. Tangible

This final dimension refers to the effects on customers that the tools and all of the tangible elements comprising the physical context of the service encounter can have. These elements may include equipment, physical facilities, personnel or communication media. These elements, known as "service landscapes" (servicescapes), can influence customers in physiological, cognitive and emotional ways (Bitner 1992). Therefore, when African importers meet a new wine supplier with limited reputation, they naturally rely on these elements of the servicescape, such as its wine storage site, website or offices. Some Africans have described this as a fundamental preliminary step before initiating a relationship. We thus propose hypothesis H3.

H3.– A wine importer's confidence in a wine supplier is positively related to the five dimensions of overall service quality.

15.3.2.2. *Perceived value*

When a buyer and supplier engage in a lasting relationship, it is to work in a way that adds value for both parties. However, the common definition of perceived value is typically seen as a compromise between the benefits the customer receives and the sacrifices made during the interaction with the seller. This simplified definition does not encompass all elements that can make value a precursor to trust (Zeithaml 1988, p. 14). It is essential to consider the social and emotional components that are vital for building trust (Eggert et al. 2006). Doney et al. (2007) have established a connection between trust building and the perceived value of the provider's services. Therefore, we will propose hypothesis H4.

H4.– An importer's confidence in their supplier is positively related to the perceived value of the service offer.

15.4. Commitment to buyer–supplier relationships

Trust and commitment have been described as critical antecedents of inter-organizational collaboration (Fan et al. 2012). In fact, while mutual commitment plays an important role in establishing trust in inter-business relationships, trust, in turn, reduces the perceived risk in the relationship and leads to commitment. In this context, if a supplier is not perceived as competent and honest enough, the customer may not demonstrate commitment to the relationship (Morgan and Hunt 1994). Therefore, we propose hypothesis H5.

H5.– Trust in an African wine importer leads to commitment in a wine supplier.

15.5. Word-of-mouth in buyer–supplier relationships

Word-of-mouth can be described as positive comments made by individuals who have experienced products or services, directed towards those who wish to try these products or services (Zhao and Xie 2011). In B2B contexts, the influence of word-of-mouth on customer purchasing decisions has been widely acknowledged as significant (Khuong et al. 2021). Interpersonal communications in inter-business relationships lead to positive word-of-mouth (Huotari et al. 2015). Additionally, empathy and problem-solving abilities (Hébert and Hauf 2015) also contribute to a positive perception in this type of relationship. We thus propose hypothesis H6.

H6.– An African importer's trust in a wine supplier leads to positive word-of-mouth.

15.6. Problem and conceptual model

Our primary research question pertains to the impact of social antecedents and supplier-related antecedents on the construction of trust in a B2B relationship. The model was inspired by Doney et al. (2007), where they outline the factors related to supplier behavior and offerings for establishing trust in B2B.

Based on this, the model for the current research was developed (see Figure 15.2). Customer orientation was not retained, as the perception of the supplier is not considered in this study. In contrast, given the context of the wine industry in Africa, where wine culture is currently relatively weak, recommendations from peers appear to be relevant. Therefore, this variable has been included as a consequence of trust.

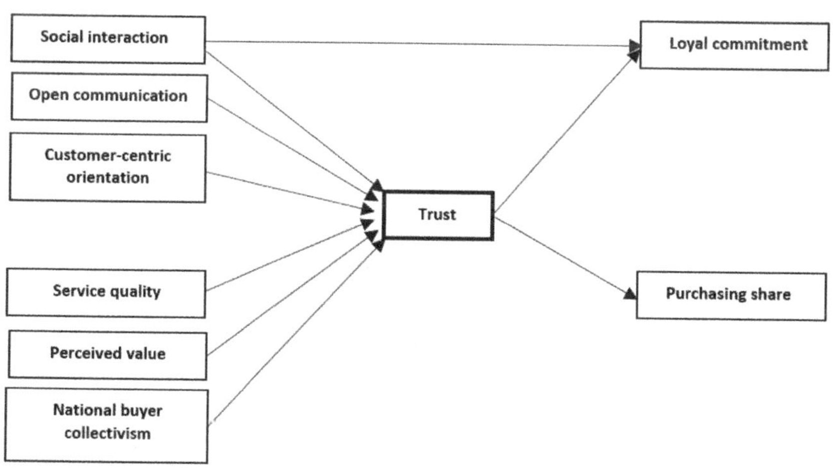

Figure 15.1. *Conceptual model of Doney et al. (2007)*

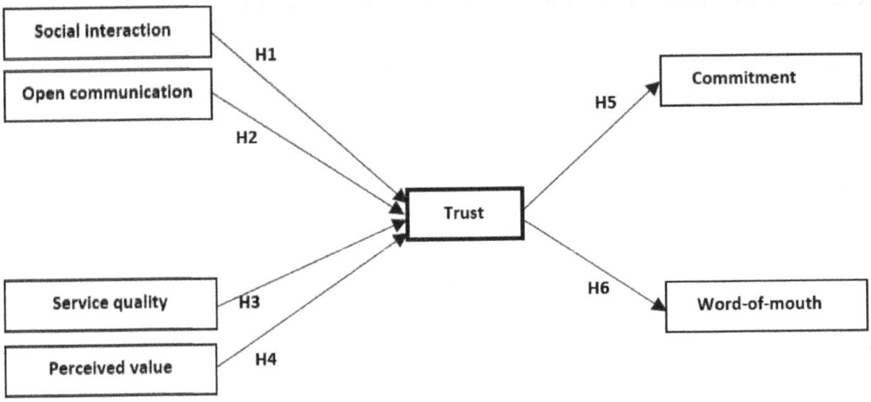

Figure 15.2. *Conceptual model for commitment, relationship and trust*

15.7. Research methods and preliminary results

Our approach encompasses both qualitative and quantitative methods. The respondents (African importers) were divided into two groups based on their origin, whether they were from a French-speaking African country (PAF) or an English-speaking African country (PAA).

PAF	Cameroon/Côte d'Ivoire/Togo/Burkina Faso/RDC (Democratic Republic of Congo)/Benin/Congo
PAA	Nigeria/Kenya/Ghana/Tanzania

Table 15.2. *Grouping of countries studied by language spoken*

The average age of the subjects is approximately 40 years, and the company sizes vary from fewer than 5 to over 10 employees. Qualitative data was collected using a questionnaire and semi-structured interviews with open-ended questions. Participants were interviewed in either French or English via WhatsApp calls. These individuals were not comfortable with audio recordings, so note-taking was used during the conversations. Trust and its determinants were then deconstructed and examined through thematic analysis of the various responses obtained. The primary interviews were conducted through WhatsApp calls in either French or English, from France, between June and September 2019. Our preference for using WhatsApp was driven by the high cost of direct phone calls between France and Africa and the widespread use of this mobile application in Africa.

As for the quantitative analysis, it was carried out with a sample of 50 African companies that import wines from Europe. The participants received a questionnaire via email and WhatsApp in September 2022.

15.7.1. *Collection and sample*

The data was gathered through a questionnaire distributed via email and WhatsApp to respondents responsible for procurement in African companies importing wine from Europe to Africa. Data collection took place during the period from September 1 to September 30, 2022.

The respondents were either business owners or individuals in roles related to purchasing or wine supplier management within their companies. They were chosen from a prospecting database compiled between 2019 and 2022. Participation in the study was voluntary. All participants were guaranteed that their data would be anonymized, and the collected responses would be analyzed solely for research purposes. The online survey was conducted in French for respondents from Francophone countries and in English for respondents from Anglophone countries.

The respondents were requested to assess the survey questions from the standpoint of a buyer and their association with their primary wine supplier. Respondents originated from either companies exclusively importing wine (30%) or companies importing a range of food products (70%).

The total sample comprises 40 professionals, predominantly male (92.5% men). The average age of the respondents is precisely 39.39 years (standard deviation: 8.47). The average number of years of experience within the current organization is 5.53 years (standard deviation: 1.34), while the average number of years of experience with the primary supplier is 3.10 years (standard deviation: 1.25).

15.7.2. *Psychometric measurements and qualities*

The scales chosen for this study were adapted from existing scales and used in a simplified Likert format ranging from 1 (strongly disagree) to 5 (strongly agree).

Our results also indicate good psychometric properties. The social interaction scale (ρ: 0.830; ρ_{vc}: 0.711) was adapted from Damperat and Jolibert (2009); open communication (ρ: 0.800; ρ_{vc}: 0.505) was adapted from Wiengarten et al. (2010). The service quality (second order, ρ: 0.890; ρ_{vc}: 0.624) was measured using the SERVPERF tool. However, due to its length (22 items), we used a shorter version as proposed by Albouy et al. (2019), consisting of two items per dimension, making a total of 10 items. The perceived value (ρ: 0.828; ρ_{vc}: 0.546) was adapted from Hansen et al. (2008), word-of-mouth (ρ: 0.872; ρ_{vc}: 0.777) from Kumar et al. (1994) and commitment (ρ: 0.807; ρ_{vc}: 0.512) from Morgan and Hunt (1994). Finally, the trust scale (ρ: 0.773; ρ_{vc}: 0.540) was adapted from Rašković and Mörec (2013).

15.7.3. *Hypothesis testing*

The hypotheses were tested using AMOS 27 software. The results of the direct relationships are summarized in Table 15.3.

Relationships	Main effect (β)	t	p-value
Social interaction→ trust	0.022	0.136	0.893
Open communication → trust	0.142	0.748	0.459
Quality of service→ trust	0.426	2.183	0.036
Perceived value→ trust	0.445	2.278	0.029
Trust → commitment	0.638	5.102	0.000
Trust → word-of-mouth	0.643	5.180	0.000

Table 15.3. *Hypothesis test results*

Hypothesis testing suggests that not all trust antecedents have a significant effect. This is particularly evident in the case of communication-related antecedents (social interaction and open communication), which seem to have no effect. This may also be due to the small sample size, which may not have provided enough degrees of freedom for analysis. A bootstrap procedure (n = 2,000) was carried out, but it did not alter the conclusions regarding the hypotheses. However, despite the sample size, significant relationships can be observed between the quality of service (β: 0.426; $p < 0.036$) and perceived value (β: 0.445; $p < 0.029$). Finally, the hypotheses concerning the impact of trust on engagement (β: 0.638; $p < 0.000$) and word-of-mouth (β: 0.643; $p < 0.000$) are well supported by the collected data.

Finally, as demonstrated in Table 15.4, the exploration of indirect effects allows us to conclude the existence of the impact of service quality and perceived value on commitment and word-of-mouth through trust, in line with existing literature.

Relationships	Indirect effect	Z (Sobel)	p-value
Quality → trust → commitment	0.272	2.007	0.0223
Quality → trust → word-of-mouth	0.274	2.012	0.0221
Value→ trust → commitment	0.284	2.080	0.0187
Value→ trust → word-of-mouth	0.286	2.085	0.0185

Table 15.4. *Significant indirect effects*

15.8. Discussion

This study examined the impact of social antecedents (such as the frequency of social interaction and open communication) and offer-related antecedents (service quality, perceived value, recommendations from relatives) on the development of trust between a wine importer and a supplier. Additionally, it aimed to determine whether this could lead to a relational commitment.

Drawing from our results, the study suggests a positive relationship between a high frequency of interaction and trust (H1). The study also posits that quality service fosters trust between African importers and suppliers (H3). However, open communication has no significant impact on trust-building (H2), whereas the value of the provider's service offering is assumed to generate trust (H4). Finally, relational commitment is assumed to be directly associated with the prior establishment of trust (H5).

Results from the qualitative study validate and extend the existing academic literature on building business-to-business trust and are in line with the results obtained during the qualitative analysis.

Therefore, the study's results suggest that the quality of service plays a significant role in trust formation. Consequently, it implies that wine suppliers should invest in the type of service they offer to African customers, as this profoundly influences trust building. During the interviews, several importers mentioned the frequent mismatch between their needs and the services offered by their European wine suppliers. The exchange of high-quality information alone is insufficient for building trust. Instead, the value of the service offerings is more appealing to Africans. Companies should invest in developing service offerings tailored to their African partners, as this strengthens trust within the supply chain. It is worth noting that Prahalad (2004, p. 19) suggests that Western companies should overcome their biases (in our study, biases pertain to the African continent) by providing affordable offers without compromising on quality or performance associated with such dedicated offerings.

Furthermore, it has been observed that a strong interaction between the buyer and the supplier positively influences trust and commitment, and interviews have revealed that African importers place great importance on warm, familiar and regular exchanges with their counterparts. Therefore, our results recommend that companies aiming to establish long-term collaboration with African importers should invest in implementing a strategy of regular visits to Africa to facilitate buyer engagement in the relationship. We believe that these visits are even more essential in Africa, as professional relationships may, perhaps more than elsewhere, lead to more personal connections related to family or religion.

Another significant outcome arising from this study is the positive influence of trust on emotional commitment in the relationship. Beyond trust, it is suggested that emotional attachment could enhance the development of a long-term business-to-business relationship.

15.9. Managerial implications

The study's results carry several implications for managers. The first implication pertains to increasing budgets for frequent trips to Africa to meet with customers. Suppliers should make an effort to engage with African importers more regularly, even if it is not necessarily for business discussions. Attention should also be given to the quality of services offered, avoiding the perception that African importers are receiving discounted services (Duquesnois et al. 2020). The trust that arises from this evolves over time towards greater buyer commitment and long-term relationships.

The availability and user-friendliness of information tools like WhatsApp can significantly enhance information exchange between companies, especially considering that emails are less frequently used in Africa compared to Europe. Furthermore, recognizing that commitment can have long-term implications for the sustainability of a relationship, it is crucial for companies to establish significant dedicated relationship roles for customers in Africa when building trust with partner companies. As an example, we can point to a highly skilled player in the wine market development in Africa who, during the Covid-19 period, resigned from his position because he felt he could no longer fulfill his crucial relational role in Africa for the smooth operation of his business. According to this individual, remote communication tools such as Zoom or Teams alter the unique and, in his opinion, essential relationships necessary for successful business operations in Africa.

15.10. Conclusion

The understanding of the nature of business-to-business relationships and their evolution is a subject of investigation in various sectors, particularly in international trade. In a context where access to information is increasingly available, especially in the wine industry, which is characterized by ongoing competition in various markets, sustainable inter-organizational relationships can be achieved through frequent social interaction between the leaders of partner organizations. This social interaction should be complemented by a high-quality service offering from the supplier. This study provides relevant insights that lead us to consider that inter-organizational communication is a highly important, even critical, relational skill that can lead to mutual benefits within the collaboration between buyers and suppliers.

The findings of this study align with the literature in various business sectors, emphasizing the significance of positive social interaction in influencing trust. Embracing a transdisciplinary approach, the study's results underscore the importance of understanding "inter-organizational" communication from a more relational perspective rather than solely focusing on material and financial transactional exchanges.

Consequently, this study highlights the essential foundations for developing distinctive (in the sense of being difficult to replicate) relational communication skills for effective B2B relationships in Africa. Furthermore, these distinctive capabilities lead us to consider specific practices that take into account the cultural context in which this inter-organizational communication occurs. For instance, our study prompts us to contemplate the intentional use of religious vocabulary in this inter-organizational communication, which was highly valued by African importers during our qualitative analysis. We acknowledge that such a practice may be less feasible, for instance, in France.

15.11. Appendix: questionnaire

Variables
Social interaction (Damperat and Jolibert 2009) (1) My primary supplier calls me regularly to talk. (2) My primary supplier visits me regularly. ***Open communication*** (Wiengarten et al. 2010) (1) My primary supplier informs me about long-term strategic plans and events, such as: entering new markets or acquiring new customers. (2) My primary supplier keeps me informed about new products or changes in existing products. (3) Information exchanged with my primary supplier has added value for the needs of my business. (4) The information exchanged with my primary supplier is reliable for our business needs. ***Quality of service*** (Albouy et al. 2019) *Helpfulness* (1) My primary supplier's staff provides us with prompt service. (2) My primary supplier is available to respond to my requests. *Reliability* (3) My primary supplier is trustworthy. (4) If we encounter a problem, the staff of my primary supplier has a friendly and reassuring attitude. *Assurance* (5) With my primary supplier, I securely make my payments. (6) I can trust my primary supplier's staff. *Tangible* (8) The layout of my primary supplier's site is visually appealing. (9) The appearance of my primary supplier's physical facilities is consistent with the type of service provided. *Empathy* (10) My primary supplier's staff is paying attention to us. *Individual* (11) My primary supplier seems to us to be acting in our best interests.

Perceived value (Hansen et al. 2008)

(1) My relationship with this supplier is very beneficial to me.

(2) Our costs have decreased more than we thought when the relationship with this supplier was established.

(3) It is more interesting for us to do business with this supplier than with others.

(4) As a customer of this supplier, I get more for my money.

Word-of-mouth (Kumar et al. 1994)

(1) I will probably encourage other acquaintances to do business with this supplier.

(2) I will recommend this supplier to someone who asks me for advice.

Commitment (Morgan and Hunt 1994)

(1) Even if I could, I wouldn't let my primary supplier down because I like being associated with them.

(2) I want to remain a member of my primary supplier's network because I really value the relationship with them.

(3) My positive impressions of my primary supplier are a major reason to continue working with them.

(4) I hope that my relationship with my main supplier will continue in the long term.

Trust (Rašković and Mörec 2013)

(1) My primary supplier is known for honesty.

(2) My primary supplier keeps their word.

(3) My primary supplier is seeking the benefit of their partner.

(4) One of the characteristics of my relationship with my primary supplier is that no one is expected to make a request to the other that might be disadvantageous to them.

(5) I can always rely on my primary supplier when needed.

(6) I believe my primary supplier will work in the future to maintain a good relationship with my business.

15.12. References

Ahearne, M., Hall, Z., Krishnamurthy, P., Pourmasoudi, M. (2019). Selling in the digital age. Working Paper Series Report, Marketing Science Institute, 19–123.

Albouy, J., Galan, J.P., Maubisson, L. (2019). Building satisfaction and loyalty during a musical benefit event: A consumer-based approach. *Nonprofit and Voluntary Sector Quarterly*, 48(6), 1250–1271.

Anderson, E. and Weitz, B. (1989). Determinants of continuity in conventional industrial channel dyads. *Marketing Science*, 8(4), 310–323.

Atef, Y., Ahearne, M., Hohenberg, S., Hall, Z., Zettelmeyer, F. (2020). Open negotiation: The back-end benefits of salespeople's transparency in the front end. *Journal of Marketing Research*, 57(6), 1076–1094.

Bitner, M.J. (1992). Servicescapes: The impact of physical surroundings on customers and employees. *Journal of Marketing*, 56(2), 57–71.

Chu, S.-Y. and Fang, W.-C. (2006). Exploring the relationships of trust and commitment in supply chain management. *Journal of American Academy of Business*, 9(1), 224–228.

Damperat, M. and Jolibert, A. (2009). A dialectical model of buyer-seller relationships. *Journal of Business and Industrial Marketing*, 24(3–4), 207–217.

Donada, C. and Nogatchewsky, G. (2007). La confiance dans les relations interentreprises : une revue des recherches quantitatives. *Revue française de gestion*, 175, 111–124.

Doney, P.M. and Cannon, J.P. (1997). An examination of the nature of trust in buyer-seller relationships. *Journal of Marketing*, 61, 35–51.

Doney, P.M., Barry, J.M., Abratt, R. (2007). Trust determinants and outcomes in global B2B services. *European Journal of Marketing*, 41(9/10), 1096–1116.

Duquesnois F., Voisin V., Dikmen L. (2020). Creation and development of markets by Les Grands Chais de France in Africa. In *Wine Management and Marketing: Opportunities For Companies and Challenges for the Industry*, Cheriet, F., Maurel, C., Amadieu, P., Hannin, H. (eds). ISTE Ltd, London, and John Wiley & Sons, New York.

Dyer, J.H. and Chu, W. (2000). The determinants of trust in supplier-automaker relationships in the US, Japan and Korea. *Journal of International Business Studies*, 31(2), 259–285.

Eggert, A., Ulaga, W., Schultz, F. (2006). Value creation in the relationship life cycle: A quasi-longitudinal analysis. *Industrial Marketing Management*, 35(1), 20–27.

Fan, Y.H., Woodbine, G., Scully, G. (2012). Guanxi and its influence on the judgments of Chinese auditors. *Asia Pacific Business Review*, 18(1), 83–97.

Gao, T., Joseph Sirgy, M., Bird, M.M. (2005). Reducing buyer decision-making uncertainty in organizational purchaing: Can supplier trust, commitment, and dependence help? *Journal of Business Research*, 58(4), 397–405.

Gounaris, S.P. (2005). Internal-market orientation and its measurement. *Journal of Business Research*, 59(4), 432–448.

Hansen, H., Samuelsen, B.M., Silseth, P.R. (2008). Customer perceived value in B2B service relationships: Investigating the importance of corporate reputation. *Industrial Marketing Management*, 37(2), 206–217.

Hébert, A. and Hauf, P. (2015). Student learning through service learning: Effects on academic development, civic responsibility, interpersonal skills and practical skills. *Active Learning in Higher Education*, 16(1), 37–49.

Hofstede, G.H. (2001). *Culture's Consequences: Comparing Values, Behaviors, Institutions, and Organizations across Nations*, 2nd edition. Sage Publications, Thousand Oaks.

Huotari, L., Ulkuniemi, P., Saraniemi, S., Mäläskä, M. (2015). Analysis of content creation in social media by B2B companies. *Journal of Business & Industrial Marketing*, 30(6), 761–770.

Khuong, M., Nhi, N., Linh, L. (2021). Antecedents of word-of-mouth towards professional B2B service. *Management Science Letters*, 11(4), 1253–1266.

Kumar, N., Hibbard, J.D., Stern, L.W. (1994). *The Nature and Consequences of Marketing Channel Intermediary Commitment*. Marketing Science Institute, Cambridge, MA.

Kumar, N., Scheer, L.K., Steenkamp, J.-B.E.M. (1995). The effects of perceived interdependence on dealer attitudes. *Journal of Marketing Research*, 32(3), 348–356.

Lee, D.Y. and Dawes P.L. (2005). Guanxi, trust, and long-term orientation in Chinese business markets. *Journal of International Marketing*, 13(2), 28–56.

Leke, A., Chironga, M., Desvaux, G. (2018). *Africa's Business Revolution*. Harvard Business Press, Boston.

Lui, S.S., Ngo, H.-Y., Hon, A.H.Y. (2006). Coercive strategy in interfirm cooperation: Mediating roles of interpersonal and interorganizational trust. *Journal of Business Research*, 59(4), 466–474.

MacMillan, K., Money, K., Money, A., Downing, S. (2005). Relationship marketing in the non-profit sector: An extension and application of the commitment-trust theory. *Journal of Business Research*, 58(6), 806–818.

Meyer, K.E., Estrin, S., Bhaumik, S.K., Peng, M.W. (2009). Institutions, resources, and entry strategies in emerging economies. *Strategic Management Journal*, 30(1), 61–80.

Morgan, R.M. and Hunt, S.D. (1994). The commitment-trust theory of relationship marketing. *Journal of Marketing*, 58(3), 20–38.

Parasuraman, A., Zeithaml, V.A., Berry, L.L. (1985). A conceptual model of service quality and its implications for future research. *Journal of Marketing*, 49(4), 41–50.

Parasuraman, A., Zeithaml, V.A., Berry, L.L. (1988). SERVQUAL: A multiple-item scale for measuring consumer perceptions of service quality. *Journal of Retailing*, 64(1), 12–40.

Prahalad, C.K. (2004). *4 milliards de nouveaux consommateurs. Vaincre la pauvreté grâce au profit*. Pearson Education, Paris.

Rašković, M. and Mörec, B. (2013). Determinants of supplier-buyer relationship competitiveness in transnational companies. *Economic and Business Review*, 15(1), 5–31.

Smith, J.B. (1998). Buyer-seller relationships: Similarity, relationship management and quality. *Psychology and Marketing*, 15(1), 3–21.

Wiengarten, F., Humphreys, P., Cao, G., Fynes, B., McKittrick, A. (2010). Collaborative supply chain practices and performance: Exploring the key role of information quality. *Supply Chain Management: An International Journal*, 15(6), 463–473.

Wiertz, C., de Ruyter, K., Keen, C., Streukens, S. (2004). Cooperating for service excellence in multichannel service systems: An empirical assessment. *Journal of Business Research*, 57(4), 424–432.

Wine Institute (2020). World wine consumption [Online]. Available at: http://wineinstitute.org/.

Zaheer, A., McEvily, B., Perrone, V. (1998). Does trust matter: Exploring the effects of interorganizational and interpersonal trust on performance. *Organization Science*, 9(2), 141–159.

Zeithaml, V.A. (1988). Consumer perceptions of price, quality and value: A means-end model and synthesis of evidence. *Journal of Marketing*, 52(3), 2–22.

Zhao, M. and Xie, J. (2011). Effects of social and temporal distance on consumers' responses to peer recommendations. *Journal of Marketing Research*, 48(3), 486–496.

16

Governance and Strategy of French Wine Cooperatives and Cooperative Federations for Export

16.1. Introduction

France, the world's second-largest wine producer, with an annual output of 45–50 million hectoliters, is also the second-largest wine consumer globally. However, wine consumption has been steadily decreasing, both in terms of value and volume, since the 1960s, declining from over 100 liters per capita per year to less than 40 liters per capita per year today[1]. Consequently, despite French wines still dominating the domestic market (accounting for 65–70%), the ongoing structural decline has made export a crucial priority for French wine companies over several decades.

In this context, approximately 32% of French wines are exported (14–15 million hectoliters per year), and the growth in the value of these exports has been particularly strong for over two decades. Indeed, they have risen from €5.5 billion in 2000 to €10.6 billion in 2021[2], with almost continuous growth, except for the financial crisis of 2008–2009 and the challenges related to Covid-19 and the 2020 Trump tariffs[3], which represents a 92.7% increase over this period. Consequently, French wine and spirits exports reached €15.5 billion in the French trade balance for 2021, making it the second-largest contributor, surpassing cosmetics and trailing only behind the aeronautics industry.

Chapter written by Louis-Antoine SAISSET and Iciar PAVEZ.

1 Source: https://www.intervin.fr/etudes-et-economie-de-la-filiere/chiffres-cles.

2 Sources: DGDDI and FranceAgriMer.

3 Increase in customs duties on French wines (25%), imposed by the United States from October 2019 to March 2021, following the US/EU dispute.

On a global scale, France holds the top position as the leading wine exporter in terms of value and ranks third in terms of volume (FranceAgriMer 2022). This leading position is attributed to the high-end positioning of wines marketed for export: AOP (Appellation d'Origine Protégée) still wines, and champagne account for 49% and 28% of the total value of these exports (FranceAgriMer 2022).

However, in terms of volume, exports appear to plateau at 30–35% of French wine production, and SMEs in the French wine sector, which make up the largest number of companies, must demonstrate adaptability, agility and innovation to compete. Within this context, while wine cooperatives represent around 40% of wine production and contribute to the structuring of major wine regions, such as Champagne and Languedoc-Roussillon, it is worth exploring the situation for these collectively oriented businesses with distinct characteristics in terms of governance and economic goals. How are they evolving and shaping their export strategies? This is the subject of our study in this chapter. Utilizing information from the economic observatory of the High Council for Agricultural Cooperation (HCCA) and conducting semi-structured interviews with companies and industry organizations, we analyze the export positioning of the French wine cooperative sector and its evolution between 2015 and 2020.

This chapter is structured into three main sections: in the first section, we introduce the unique characteristics of agricultural and wine cooperatives and their significance in the French wine industry; in the second section, we delve into the evolution, downstream strategies and governance modes of wine cooperatives; in the third and final section, we provide an analysis of the export strategies of French wine cooperatives and cooperative federations.

16.2. Wine cooperatives: specificities and weight in the French wine sector

16.2.1. *Specificities of agricultural and wine cooperatives*

According to Mertens and Marée (2010), social economy enterprises, including cooperatives, are distinguished by a consistent set of values, which serve as guiding principles for their leaders and differentiate them from more conventional forms of businesses. These authors emphasize four core values: solidarity, equality, freedom and a sense of responsibility.

For the International Cooperative Alliance (ICA), which unites and represents all types of cooperatives worldwide, these organizations are primarily built on six fundamental values: mutual aid, personal responsibility, democracy, equality, equity and solidarity. In the tradition of their founders, cooperative members embrace an ethical framework grounded in honesty, transparency, social responsibility and altruism.

These strong and unique humanistic values already motivated the founding members of the Rochdale Society of Equitable Pioneers. Established on August 15, 1844, by 28 weavers from the outskirts of Manchester, it is regarded as the very first cooperative that laid the groundwork for cooperative principles, which largely persist today (Saïsset 2016).

The international cooperative principles have been revised three times since 1895, with the most recent version dating back to 1995. These principles serve to distinguish these organizations from conventional capital-based companies; there are seven of them, as outlined in Box 16.1.

– Voluntary and open membership

– Democratic member control

– Member economic participation

– Autonomy and independence

– Education, training, and information

– Cooperation among cooperatives

– Concern for community

Box 16.1. *Cooperative principles of the ICA (source: ICA statement (1995))*

The principles of "open membership" (voluntary and open to anyone interested in the cooperative's activities) and "one person, one vote" (democratic control with voting rights not proportional to capital ownership) are among the most important, as is the principle regarding the required engagement in the cooperative's activities (buying or selling through the cooperative) for each member (economic participation).

However, these principles are not static and have been implemented differently, depending on the countries and sectors of activity, and even adapted or relaxed. In France, agricultural cooperatives operate under a specific legal framework, sui generis, neither civil nor commercial, integrated into the Rural and Maritime Fisheries Code (CRPM). These organizations are defined as follows: "The purpose of agricultural cooperative societies is the collective use of any means by farmers that may facilitate or develop their economic activity, improve, or increase the results of this activity" (Article L521-1 of the CRPM).

Agricultural cooperatives are primarily designed to fulfill the professional needs of their members, rather than rewarding the capital held by their shareholders. Their members, often referred to as associate-cooperators, have a dual commitment. They are required to make a multi-year economic commitment (a minimum of three years) involving contributions or the purchase of goods and services. Additionally, they have a financial commitment, which involves subscribing to share capital proportionate to their level of activity. As a result, they play the roles of both suppliers or customers and shareholders. This dual role and commitment provide agricultural collection-sale cooperatives (involved in the purchase of agricultural products, storage, packaging and the sale of processed or unprocessed products) with a dual objective: ensuring optimal remuneration for members while consolidating financial resources at the cooperative enterprise level. This creates a genuine "cooperative dilemma" that can lead to tensions within decision-making bodies, particularly in wine cooperatives (Saïsset 2020), and can impact their governance approach.

In addition to this unique point, we can also observe that the governance of cooperatives significantly differs from that of capitalist enterprises due to its democratic nature (each member having one vote, and only one, in general meetings), and the dual status of associate-cooperators. In this context, the economic commitment takes precedence over the financial commitment, with a minimum contribution period of three years and a limitation of interest to shares.

Furthermore, the strong local presence of agricultural and wine cooperatives (Filippi 2013), which can only engage with associate-cooperators whose farms or plots are located within the "territorial district" specified in their bylaws (Article L521-2 of the CRPM), has an impact on the decision-making processes and strategies of these unique enterprises.

16.2.2. *Importance of wine cooperatives*

France is home to 2,200 agricultural cooperatives (excluding cooperatives for shared agricultural equipment) with membership from three-quarters of farmers (approximately 292,000 farms). These cooperatives generate a consolidated turnover of €86 billion, which also includes the revenues of cooperative subsidiaries (La Coopération Agricole 2021). They account for 40% of the agri-food sector.

In this highly concentrated landscape, the wine sector comprises a significant number of small and medium-sized cooperatives: 570 wine cooperatives (first tier) and 31 cooperative federations (second tier) in 2018, making up 27% of all agricultural cooperative enterprises. Their combined turnover is estimated at €5.1 billion[4], accounting for only 6% of the total turnover.

570 wine cooperatives, 31 marketing cooperative federations, 35,000–40,000 members (wine grower members), 17,000 employees, 300,000 hectares, 15–20 million hectoliters of production and €5.1 billion in turnover.

Top 10 wine cooperatives

Cordier by InVivo (all of France), Nicolas Feuillatte (Champagne), Alliance Fine Champagne (Champagne), Union Champagne (Champagne), Terre de Vignerons (Bordeaux), Union Auboise (Champagne), Cellier des Dauphins (Vallée du Rhône), Les Vignerons de Tutiac (Bordeaux), Producteurs Plaimont (South-West), COGEVI (Champagne).

Some important wine cooperative brands

Cellier des Dauphins, Colombelle, Fruité Catalan, Jacquart, Jaillance, Nicolas Feuillate, Rauzan, Remy Pannier, Veuve Devaux, Wolfberger.

Share of wine cooperatives' production

37% of the total volume. Of which PDO: 38%; PGI: 69%; other wines: 30%.

Box 16.2. *Economic weight of wine cooperatives in France (sources: HCCA, La Coopération Agricole-Vignerons Coopérateurs de France)*

Wine cooperatives play a significant role in the French wine industry. As indicated in Box 16.2, they encompass approximately 60% of the existing wine growers (which numbered 70,000 in 2010 and 59,000 in 2020). However, it is quite challenging to determine the exact number of wine grower members due to the possibility of joining multiple cooperatives and being partial contributors. These organizations account for just under 40% of total wine production, with a more pronounced presence in the Protected Geographical Indications (PGI), accounting for nearly 70%, primarily due to varietal wines.

Furthermore, wine cooperatives employ 17,000 workers, constituting 22% of the direct jobs within the French wine industry (totaling 77,000 jobs). Additionally, they contribute to 25% of the sector's overall turnover (FranceAgriMer 2014), giving them a substantial economic influence.

4 Sources: HCCA and (Pavez et al. 2021).

16.3. Developments in the French wine cooperative sector: downstream strategies and governance modes

16.3.1. *Merger and consolidation of French wine cooperatives*

After reaching a peak during the 1970s, the number of wine cooperatives has continued to decline, especially in the south of France, due to various overproduction and market adaptation crises that led to a significant reduction in production areas. Various grubbing-up premium schemes were initially temporary in the 1970s–1980s, and then became permanent; they accompanied these changes (Saïsset 2014), along with numerous mergers and acquisitions (Saïsset and Cheriet 2012).

Since the early 1990s, the number of wine cooperatives has dropped by nearly 46%, decreasing from 1,040 in 1991 to 570 in 2018 (see Figure 16.1). This decline follows the same trend as that of all agricultural cooperatives (-49% between 1995 and 2022) (La Coopération Agricole 2021), but it is much more significant than the decrease in vineyard acreage, which was only -17.4% over the same period. Therefore, we are witnessing a more pronounced concentration phenomenon compared to the overall reduction in vineyard acreage.

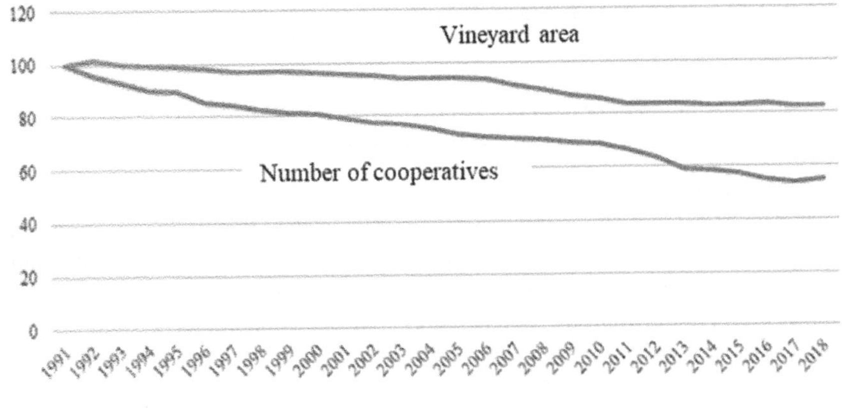

Figure 16.1. *Evolution of the number of French wine cooperatives and total vineyard area in France (base index 100 in 1991) (sources: DGDDI, CCVF and HCCA). For a color version of this figure, see www.iste.co.uk/cheriet/wine2.zip*

If we consider that at least 80% of cooperative disappearances result from a merger between two cooperatives (one absorbing, one absorbed), it can be suggested that there have been over 350 mergers during the period under consideration, averaging 14 mergers per year, with more pronounced waves in the early 1990s and mid-2000s (a severe wine crisis between 2004 and 2011 with a significant drop in prices).

These concentration phenomena through mergers have been studied in the Bordeaux (Bélis-Bergouignan and Corade 2008; Corade and Lacour 2015) and Languedoc (Saïsset et al. 2017) vineyards, emphasizing their significance and complexity, along with their undeniable territorial impact. Roth and Menguy (2015) demonstrate that strategic mergers in agricultural cooperatives are driven by critical size issues, resource pooling and competitiveness. In the wine sector, they often aim to counter the decline in acreage and production volumes or address significant financial challenges. Hence, the majority of these mergers can be termed as "defensive mergers" in response to crisis situations (Saïsset and Cheriet 2012).

This "race for size" might lead us to believe that wine cooperatives are highly concentrated and relatively large. However, this is not the case: the top 20 represents only 30.7% of the total[5] turnover, while the top 75 agricultural[6] cooperatives make up 82.1% of the turnover. Additionally, the average turnover of wine cooperatives in 2018 was €6.8 million[7], compared to €35 million for all agricultural cooperatives. Furthermore, the average production of a cooperative winery remains relatively stable over time, ranging between 35,000 and 40,000 hectoliters.

16.3.2. *Integration movement downstream and evolution of governance modes in the French wine cooperative sector*

The external growth phenomenon characterized by horizontal integration through mergers is complemented by a movement towards downstream integration by cooperatives. Beyond achieving economies of scale or maintaining production volumes, these cooperatives aim to create value by enhancing their commercial positioning, developing brands and expanding their exports. In the broader agricultural cooperative sector, Filippi and Triboulet (2011) demonstrate that this downstream integration involves various forms of strategic alliances between cooperative groups.

In reality, there are several phases related to consolidation and alliances within the wine cooperative sector, as analyzed by Saïsset (2014, 2016) and Corade and Lacour (2015). Within this framework, the establishment of cooperative federations follows the development of "original" or "autonomous" cooperatives, as described by Saïsset (2016) and Corade and Lacour (2015). These original cooperatives maintain direct relationships with their members without any satellite structures. In these traditional first-level cooperatives, governance is relatively centralized, with limited involvement of external shareholders (non-cooperator associates). The

5 Source: HCCA.

6 *Idem.*

7 Source: La Coopération Agricole.

cooperative federations, which bring together at least two cooperatives, were created on a large scale during the 1970s and 1980s, aided by the government to develop "producer groups" aimed at improving wine quality (providing technical support and restructuring the vineyards) and consolidating the supply to enhance competitiveness (Bartoli et al. 1987). Cooperative federations can serve as an intermediate stage before a merger or be an independent objective, often driven by commercial purposes (Corade and Lacour 2015). These second-level structures typically feature a more decentralized and indirect form of governance in comparison to the winemaker-cooperatives; they may be more open to non-cooperative shareholders, including capitalist enterprises that can hold shares without necessitating changes to the statutes.

The subsidiaries of agricultural cooperatives have primarily expanded since the 1990s, thanks to Law 91-5 of January 3, 1991, aimed at diversifying activities or extending them downstream, either exclusively (100% subsidiary) or in collaboration with other companies (acquisition of a majority stake). According to Filippi and Triboulet (2011), these more modern and flexible forms of strategic alliances have gradually taken precedence over cooperative federations and even SICA (agricultural collective interest companies) that previously dominated the landscape. In fact, the number of cooperative federations in the wine sector was significantly reduced between 1990 and 2020, primarily due to governance issues related to the dilution of property rights and a lack of control in the board of directors (Saïsset 2014). These subsidiary companies are predominantly organized under the legal structure of an SAS (simplified joint-stock company), known for its flexibility and adaptability. Their ownership structure is more open compared to cooperative federations or SICA, and their governance model is even more decentralized and sometimes quite distinct from the parent cooperative company.

The proliferation of cooperative federations, SICAs and subsidiaries under traditional commercial law has led to the formation of wine cooperative groups. Prominent examples include the expansion of Nicolas Feuillate, who recently, after merging with the Regional Cooperative of Champagne, became Wines, Terroirs and Vignerons de Champagne (turnover of €300 million), and the establishment of the national leader Cordier by InVivo (turnover of €400 million) through a merger between Vinadéis and InVivo Wine. The governance of these groups has become more complex, involving numerous stakeholders. As for members, they face challenges in grasping the financial aspects and internationalization of their group, as highlighted by Koulytchizky and Mauget (2003) in the context of agricultural cooperation. Consequently, there is a risk that market-driven considerations might overshadow the cooperative principles (Deroy and Thénot 2015). A governance approach that is more cognitive and partnership-oriented, allowing members and administrators to regain control of their group while embracing sustainability and corporate social responsibility initiatives, becomes essential (Saïsset 2016).

16.4. Export strategies of French wine cooperatives and wine cooperative federations

16.4.1. *Study methodology*

The methodology of our study is primarily based on the analysis of data from the HCCA economic observatory, which contains (non-exhaustive) information about the sales value in France and exports of companies in the wine cooperative sector during the period 2015–2020. The companies included in the database are either cooperatives or cooperative federations. They provide their individual financial statements and consolidated financial statements to the HCCA, as applicable. We focused on using individual financial statements because the consolidated accounts, which would have allowed us to assess the export performance of wine cooperative groups, are often incomplete. The source data was provided in Excel format, and it was thoroughly reviewed, prepared and exported to the Stata 14 software. Merging the annual datasets required having at least one common identifier, which, in our case, was the SIREN number (System for Identifying Businesses, assigned by INSEE[8]). This unique identifier is specific to each company and facilitates the identification of each wine cooperative or cooperative federation. We constructed various indicators using variables from the source files, and these indicators are detailed in the subsequent analysis. It is worth noting that companies do not consistently report their export turnover accurately, as it is not essential for proper accounting practices. This introduces a bias in our study, especially considering the limited sample size, which is more significant for cooperative federations than for cooperatives.

Secondly, our methodology relies on 11 semi-structured exploratory interviews with leaders of both capital companies and French wine cooperatives, as well as representatives of organizations associated with the industry. These interviews were conducted primarily as part of the In'FAAQT Chair's[9] action program. This approach provided us with a comprehensive understanding of the subject matter. Our interview framework focused on aspects such as export experience, the company's motivations for engaging in exports, the strategies implemented to

8 INSEE: National Institute of Statistics and Economic Studies.

9 Teaching and research chair "Innovating in the agricultural and agrifood chains, quality and territories", a collaboration between L'Institut Agro Montpellier, the Purpan School of Engineering and the National School of Agronomy of Toulouse.

facilitate this activity and the challenges faced. The interviews were recorded and, for the most part, transcribed to facilitate content analysis.

16.4.2. *Share of direct wine exports by wine cooperatives and cooperative federations in French wine exports*

Between 2015 and 2020, cooperatives and cooperative federations contributed an average of 6.3% to the value of French wine exports, with a low of 5.6% in 2015 and a high of 7.4% in 2016. These percentages align with those of Legouy and Dallot (2019). On average, 3.7% of total wine exports were carried out by cooperatives, and 2.5% by cooperative federations (see Figure 16.2).

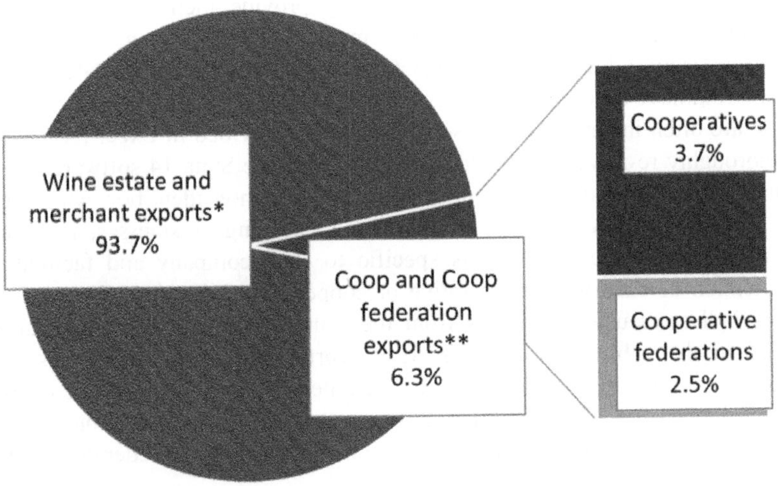

Figure 16.2. *French wine exports (% of value 2015–2020) (source: *calculation by authors based on FranceAgriMer (2021); ** HCCA data). For a color version of this figure, see www.iste.co.uk/cheriet/wine2.zip*

Table 16.1 displays the value of French wine export by wine estates and merchants, as well as by cooperatives and cooperative federations. Over the period from 2015 to 2020, the exports by wine estates and merchants, with an average value of €8.37 billion, clearly outweigh those of cooperatives and cooperative federations. The value of these exports rose from €7.82 billion in 2015 to €8.20 billion in 2020; whereas, the exports by cooperatives and cooperative federations increased from €0.46 billion in 2015 to €0.56 billion in 2020. During this period, there was a significantly positive growth, with cooperatives and cooperative federations performing better.

Figure 16.3 illustrates this trend more clearly. It reveals a 16% increase between 2015 and 2020 for cooperatives and cooperative federations; whereas, other companies only experience a 5% increase.

In both cases, 2020 represents a setback due to the combined impact of the Trump taxes and the Covid-19 pandemic.

	2015	2016	2017	2018	2019	2020	Avg.
Wine estates and merchants*	7.82	7.67	8.57	8.79	9.20	8.20	**8.37**
Cooperatives and cooperative federations **	0.46	0.62	0.54	0.58	0.60	0.54	**0.56**

Table 16.1. *Evolution of French wine export by category of companies in billions of euros (source: *calculation by the authors based on FranceAgriMer (2021); ** HCCA data)*

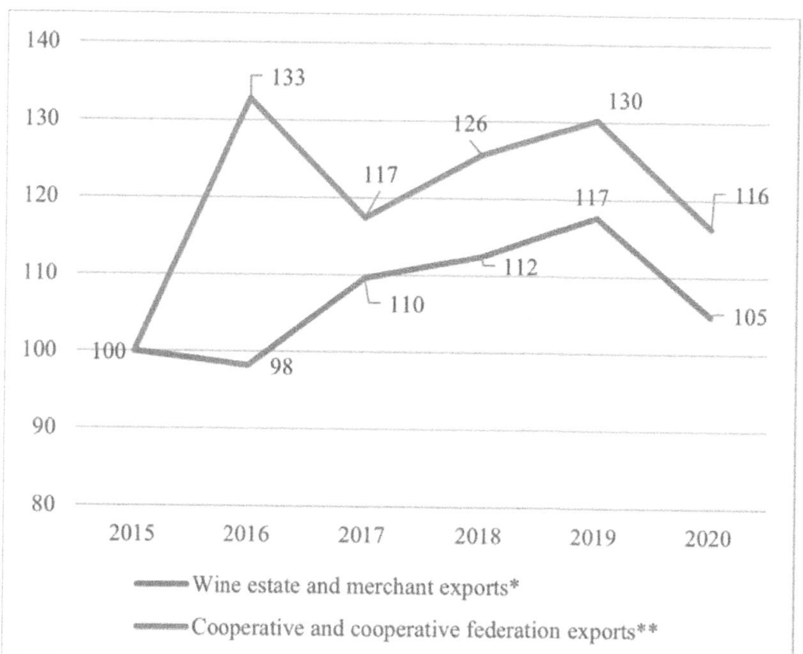

Figure 16.3. *Evolution of the value of French wine exports (base 100 in 2015) (source: *calculation based on FranceAgriMer (2021); ** HCCA data). For a color version of this figure, see www.iste.co.uk/cheriet/wine2.zip*

16.4.3. *Direct exports conducted by French wine cooperatives*

Over the period 2015–2020, there was an average of 521 wine cooperatives in France. Among them, 32.2% engaged in direct exports, which amounts to an average of 164 cooperatives. The number of cooperatives involved in direct exports increased until 2018, then decreased in 2019 and 2020. Nevertheless, when comparing the first and last years of the period, the number remained relatively stable, with slightly over 130 cooperatives.

According to our interviews with cooperative leaders, the primary motivation for venturing into exports was to mitigate the commercial risks in the French market. Most cooperatives, traditionally concentrated on the domestic market, explored foreign market opportunities through indirect exports, primarily facilitated by wine merchants. This strategy of indirect exports and a focus on the domestic market remains predominant. However, the domestic market entails escalating risks, as it is highly competitive, with the growing influence of large retailers, which act as both wine purchasers from cooperatives and competitors through their own brands.

Figure 16.4. *Number of French wine cooperatives exporting directly and the value of exports by these cooperatives (source: HCCA). For a color version of this figure, see www.iste.co.uk/cheriet/wine2.zip*

Furthermore, the French market is mature in terms of demand. These factors have exerted downward pressure on wine prices and margins. “Selling on the French market is completely off track. As long as we are dominated by the large retailers, who dictate prices, it is difficult to thrive" (Interview 2). The value of direct exports by cooperatives shows positive growth: €351.18 million in 2020, compared to €259.78 million in 2015. However, these exports declined between 2019 and 2020, influenced by the Trump taxes and the Covid-19 pandemic (see Figure 16.4). One leader explains the growth of their cooperative: "We have developed our packaging. Previously, we produced 30,000 or 40,000 bottles in total. Now, with exports, we are close to 200,000 bottles. We have implemented quality standards to enhance our products" (Interview 4).

The direct export index for the period 2015–2020 (base 100 in 2015) confirms the substantial growth for cooperatives: +43% between 2015 and 2019 (see Figure 16.5). However, there was a decline in 2020.

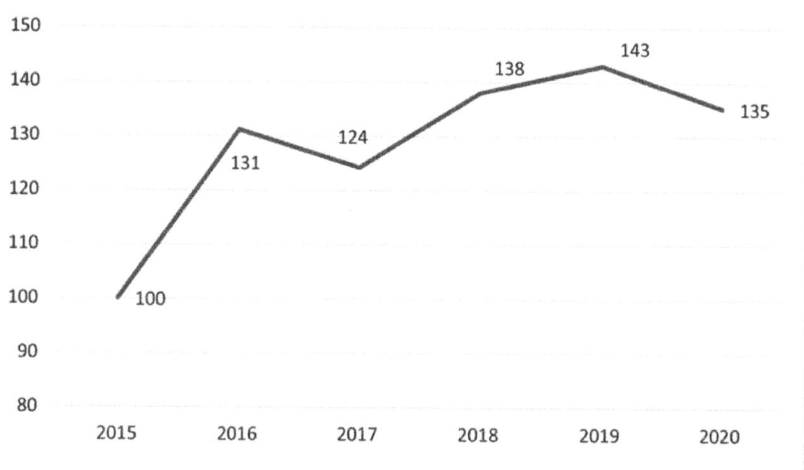

Figure 16.5. *Evolution of the direct export turnover of French wine cooperatives (base 100 in 2015) (source: HCCA. Index = (indicator year n/indicator reference year) x 100)*

To assess the significance of wine export sales, we also calculated the proportion of export turnover for wine cooperatives in relation to their total turnover. This export rate increased from 6.88% in 2015 to 11.46% in 2020 (see Figure 16.6), indicating dynamism and a growing internationalization of business channels.

Nevertheless, as mentioned before, some major risks for the wine industry have arisen from external factors since late 2019: the Trump tariffs affecting exports, along with the pandemic, which led to border closures and disruptions in both production and trade. "With this wine crisis now added to the Covid crisis... Because even before Covid, the crisis was already deep, and Covid did not make it any better" (Interview 11). One of the most significant economic impacts has been on the logistics side, resulting from increased transportation costs and delays, both during and after the pandemic. "The cost of transporting a €2,000 container has gone up to €10,000, and transport times have extended by six months" (Interview 9).

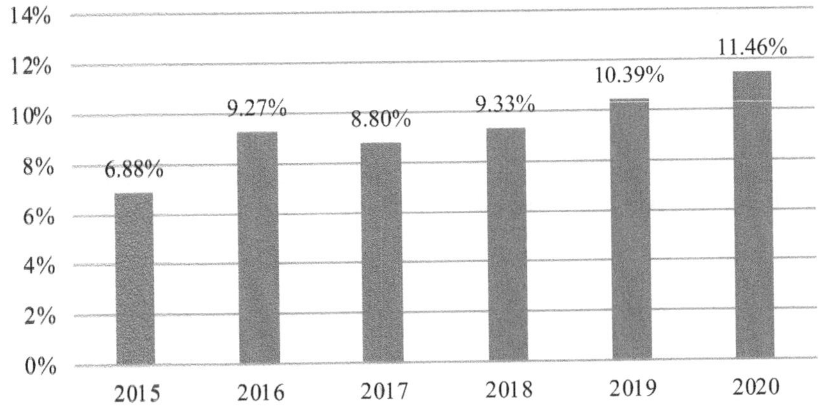

Figure 16.6. *Direct export rate (% of total turnover) of French wine cooperatives (source: HCCA)*

It should also be noted that one of the main barriers to exporting is related to the difficulty of investing in intangible assets (for example, marketing strategy, creating commercial networks and subsidiaries). According to the interviewees, "it is easier for the board of directors to approve tangible investments… But when you venture into foreign markets, it typically involves intangible investments, which wine growers tend to dislike!" (Interview 10).

Moreover, there are numerous other challenges that serve as motivations for export, including: environmental concerns ("The environmental aspect is the major challenge for future markets. We will not have a choice" (Interview 6), "Scandinavians have always been more environmentally conscious than the French… In organic products, there is more demand than supply" (Interview 2)); product quality and development, as well as organizational challenges. Interviewees emphasize: "We have established an expert jury committee consisting of sales directors, vine technicians and oenologists" (Interview 10). Developing human and financial capabilities is also essential for implementing the export strategy. The motivations of cooperatives are summarized in Box 16.3.

Risk reduction

– Reducing dependency

– Coping with the decline in wine prices and margins

Market diversification

– Product diversification

– Improved products and processes

– Brand development

Environmental and organizational challenges

– Differentiation through quality and sustainability approaches

– Human and financial capacity development

Box 16.3. *Motivations of French wine cooperatives to embark on export activity (source: qualitative interviews)*

In conclusion, it is worth noting that wine cooperatives, often averse to risk and characterized by governance focused on their members' short-term interests, tend to have limited or no export activity. However, a relatively stable core of 30% of these cooperatives is expanding their international presence to diversify their distribution channels and create additional value.

16.4.4. *Direct exports made by French wine cooperative federations*

On average, during the period 2015–2020, 52.7% of wine cooperative federations engaged in direct exports (see Figure 16.7). However, the number of cooperative federations slightly decreased from 31 companies in 2015 to 29 in 2020.

The value of direct exports by the cooperative federations has shown a tendency to grow, but erratically, between 2015 and 2019, where it increased from €204.23 million to €232.75 million. In 2020, the export activity of the cooperative federations was impacted by the decline in international wine trade, as mentioned earlier.

The direct export index of wine cooperative federations over the period 2015–2020 (see Figure 16.8) revealed a 14% growth between 2015 and 2019, marked by an exceptional 35% increase in 2016, followed by a slowdown. In 2020, the adverse effects of the international context were evident, with a substantial drop in the index, falling below its 2015 level by -8%.

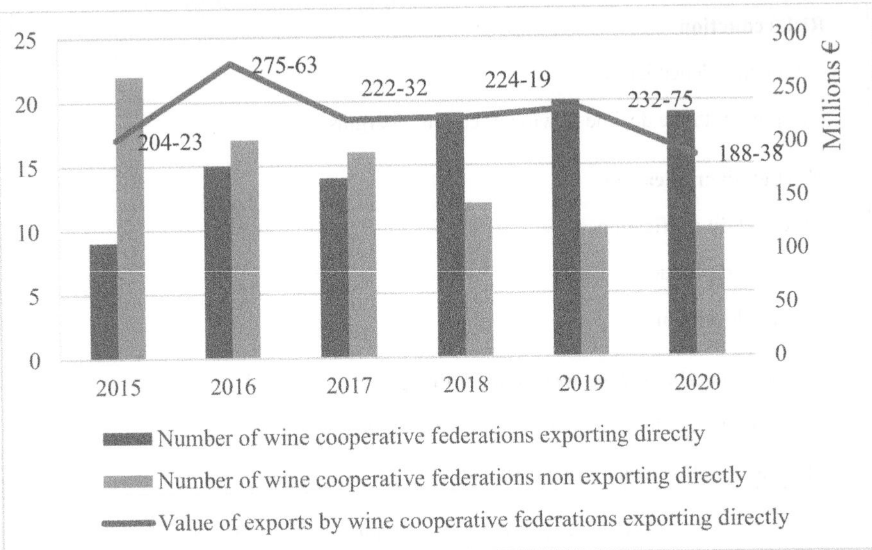

Figure 16.7. *Number of French wine cooperative federations exporting directly and the value of exports made by these cooperative federations (source: HCCA). For a color version of this figure, see www.iste.co.uk/cheriet/wine2.zip*

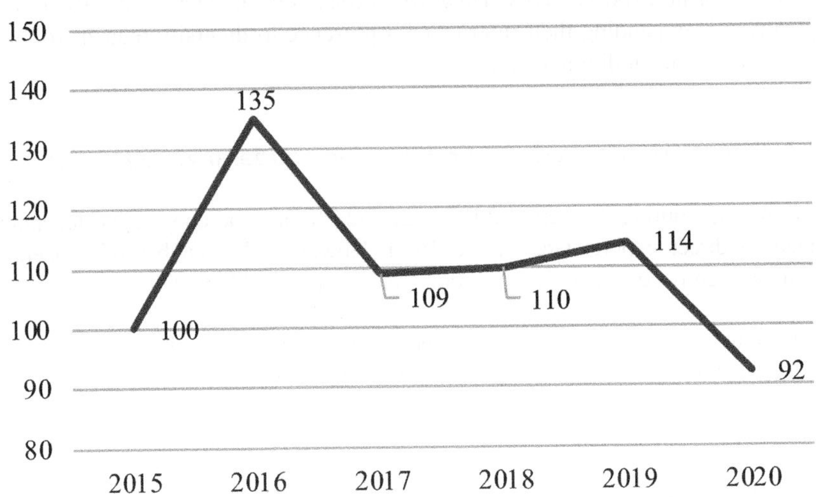

Figure 16.8. *Evolution of the direct export turnover of French wine cooperative federations (base 100 in 2015, index = (indicator year n/indicator reference year) x 100) (source: HCCA). For a color version of this figure, see www.iste.co.uk/cheriet/wine2.zip*

For cooperative federations, the average export rate over the six years studied is 22.8%, with a peak of nearly 29% in 2016 and a decline to less than 20% in 2020 (as seen in the previous crisis) (see Figure 16.9). The trend is therefore towards stability.

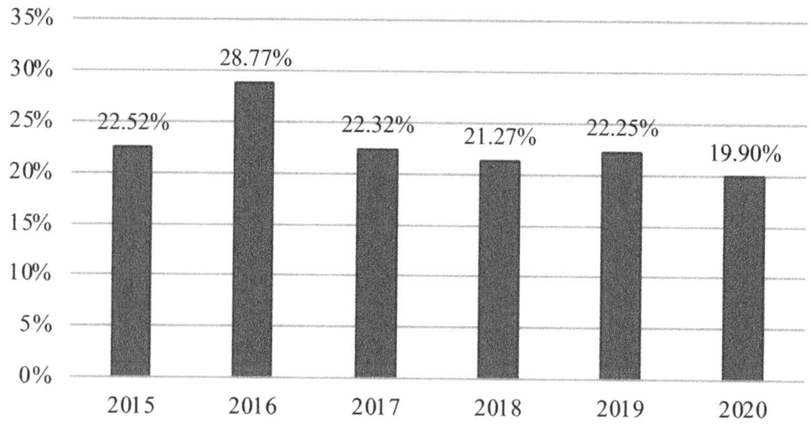

Figure 16.9. *Direct export rate (% turnover) of French wine cooperative federations (source: HCCA). For a color version of this figure, see www.iste.co.uk/cheriet/wine2.zip*

In summary, we can highlight that federations of wine cooperatives, with a more decentralized mode of governance than first-degree cooperatives and less focused on the short-term interests of wine grower members, export in greater numbers and intensity than cooperatives. However, their export activity has stagnated or even declined, indicating a degree of fragility.

16.5. Conclusion

In this chapter, we presented a comprehensive analysis of the export dynamics of French wine cooperatives and cooperative federations. Using a combination of quantitative descriptive and qualitative exploratory methodology, we shed light on the trajectories of their direct exports, which, as collective enterprises closely connected to the upstream, received limited attention within the wine industry.

In France, wine exports are heavily concentrated in the trading sector. Although wine cooperatives play a significant role in the industry, representing roughly 60% of French vineyards, these cooperatives, along with their federations, account for a relatively small portion of exports (averaging 6% between 2015 and 2020). We demonstrated that direct exports by cooperatives are on the rise, with higher growth rates than other operators, even though the overall export rate remains limited, and a substantial number of first-degree cooperatives do not engage in direct exporting. Additionally, we revealed that federations, traditionally more downstream-oriented in terms of governance and strategy, reduced their exports during the study period. However, they still maintained an export rate twice as high as first-degree cooperatives, with a more upstream-oriented mode of governance.

These findings offer opportunities for future research into the role of governance heterogeneity among wine companies in export trajectories and strategies. It would be particularly relevant to explore the subsidiaries of cooperatives and existing or emerging cooperative groups. Additionally, it would be valuable to complement our approach with an analysis of dynamics by production region and/or by country, export modes (such as importers, agents, joint ventures, etc.) and profit margins.

16.6. References

Bartoli, P., Boulet, D., Lacombe, P., Laporte, J.P., Lifran, R., Montaigne, E. (1987). *L'économie viticole française*. INRA, Paris.

Bélis-Bergouignan, M.C. and Corade, N. (2008). Fusion des coopératives vinicoles et ancrage territorial. *Revue d'Economie Regionale Urbaine*, 1, 43–68.

Corade, N. and Lacour, C. (2015) Les trajectoires d'évolution des coopératives vinicoles girondines. *Canadian Journal of Regional Science*, 38(1/3), 29–37.

Deroy, X. and Thénot, M. (2015). L'évolution des logiques coopérative et de marché dans les coopératives agricoles françaises. Le cas de Champagne Céréales. *Revue française de gestion*, 41(250), 31–47.

Filippi, M. (2013). L'ancrage territorial des entreprises coopératives agricoles et le développement durable. In *Les coopératives agricoles. Identité, gouvernance et stratégie*, Chomel, C., Declerck F., Filippi M., Mauget R., Frey O. (eds). Larcier, Brussels, 205–262.

Filippi, M. and Triboulet, P. (2011). Alliances stratégiques et formes de contrôle dans les coopératives agricoles. *Revue d'économie industrielle*, 133, 57–78.

FranceAgriMer (2014). Plan stratégique sur les perspectives de la filière viti-vinicole à l'horizon 2025. Report, Montreuil.

FranceAgriMer (2021). Vins et spiritueux, commerce extérieur. Report, Montreuil.

FranceAgriMer (2022). VIN. Fiche filière. Report, Montreuil.

Koulytchizky, S. and Mauget, R. (2003). Un siècle de développement des coopératives agricoles en France. In *Les coopératives entre territoires et mondialisation*, Touzard, J.M. and Draperi, J.F. (eds). L'Harmattan, Paris, 51–76.

La Coopération Agricole (2021). L'essentiel 2022 : coopératives agricoles et agroalimentaires.

Legouy, F. and Dallot, S. (2019). Le commerce international du vin et ses logiques territoriales à plusieurs échelles. *Géocarrefour*, 93 (93/2).

Mertens, S. and Marée, M. (2010). Les contours de l'entreprise sociale. In *La gestion des entreprises sociales*, Mertens, S. (ed.). Edipro, Liège.

Pavez, I., Saïsset, L.A., Temri, L. (2021). Gouvernance, innovation et performance des coopératives vinicoles françaises. Report, FranceAgriMer.

Roth, F. and Menguy, N. (2015). Etudier les enjeux et la performance des concentrations de structures coopératives : une approche exploratoire. *Revue de Management et de Stratégie*, 1–17.

Saïsset, L.-A. (2014). Gouvernance, investissements immatériels et gouvernance des entreprises coopératives agricoles. Le cas des coopératives vinicoles du Languedoc-Roussillon. PhD Thesis, Montpellier SupAgro, Montpellier.

Saïsset, L.-A. (2016). Les trois dimensions de la gouvernance coopérative agricole : le cas des coopératives vinicoles du Languedoc-Roussillon. *Revue Internationale de l'Economie Sociale*, 339, 19–36.

Saïsset, L.-A. (2020). Governance and performance of wine cooperatives: The case of Languedoc-Roussillon. In *Wine Management and Marketing: Opportunities for Companies and Challenges for the Industry*, Cheriet, F., Maurel, C., Amadieu, P., Hannin, H. (eds). ISTE Ltd, London, and John Wiley & Sons, New York, 161–183.

Saïsset, L.-A. and Cheriet, F. (2012). Grandir, oui, mais comment ? Analyse de la concentration par fusions des coopératives vinicoles du Languedoc-Roussillon. *Revue Internationale de l'Economie Sociale*, 326, 45–63.

Saïsset, L.-A., Cheriet, F., Couderc, J.-P. (2017). Cognitive and partnership dimensions in merger processes in agricultural cooperatives: The case of winery cooperatives in Languedoc-Roussillon. *International Journal of Entrepreneurship and Small Business*, 32(1/2), 181–207.

Conclusion

Adapting to Meet Market Challenges and Societal Expectations, and Anticipating an Uncertain Future…

While the 21st century has almost completed its first quarter, French viticulture, like in many countries, is facing several major challenges: frequent and severe climate change, increasingly radical societal expectations, logistical constraints, new distribution and consumer difficulties driven by factors such as inflation and changing consumption patterns, as well as concerning regulatory and institutional developments within the industry. In addition to these structural challenges, there are other, possibly more localized and short-term issues, such as rising raw material costs, trade competition and retaliation measures amid geopolitical crises.

The extent of these challenges should not obscure the opportunities presented to actors of the wine sector by the new national and international landscapes. This is manifested in an ongoing search for enhancing the quality of practices and products, notable innovations in processes and packaging, the rise of new markets and areas of interest for wine, new entrants and methods of marketing and consuming wine, demonstrating a robust vitality and innovation capacities in the wine industries worldwide.

This publication is therefore a valuable opportunity to explore certain aspects of these changes and the responses offered within the wine industry. The 16 chapters in this book, with diverse methodological approaches and perspectives, have tackled questions related to current challenges and future prospects for wine businesses. They aim to inform leaders and their organizations about critical issues,

---

Conclusion written by Paul AMADIEU, Hervé HANNIN, Carole MAUREL and Foued CHERIET.

necessitating individual and collective responses, sometimes of an institutional nature, to prepare the French wine sector for the challenges of the upcoming years.

These chapters frequently address the decline in average wine consumption, especially in France as well as in other countries, stringent regulations for the industry and the challenges of exporting a product with strong cultural ties. These present threats could prematurely lead to a diagnosis of decline and pessimistic forecasts for the future of this sector. Quite the contrary, these issues represent numerous challenges that can rejuvenate an industry that, deeply rooted in ancient culture and traditions, has historically demonstrated its strong capacity for innovation to adapt and meet the demands of the market and society. Viticulture has existed for millennia in various forms. It is our ambition with you, dear readers, to contemplate model(s) for the viticulture of the future, models that, without being exclusive, will address the challenges confronting viticulture and the expectations of industry stakeholders, public institutions and society.

Just like in any economic endeavor, wine companies will persist based on their ability to address not only the individual expectations of consumers and their stakeholders (employees, suppliers, financiers, local communities, etc.) but also sector-specific and societal expectations, whether individual or collective. In addition to these expectations, there are increasingly robust and intricate agricultural, climatic, logistical and administrative regulations.

Balancing all of these expectations, sometimes conflicting, may seem impossible. As a result, many winemakers at times feel overwhelmed by their (new) responsibilities, their complexity and their changing nature over time. They often entered this profession out of passion and a chosen lifestyle and are dedicated to doing their job well or, at the very least, to doing their best with the resources at their disposal, conscientiously giving their all. Nevertheless, they cannot simply claim to have fulfilled their duty, especially when society may portray them in a somewhat paradoxical light. Have winegrowers become nothing more than environmental destroyers due to their ecological footprint and producers of an addictive and harmful substance for health that ought to be banned? Are they still creators of value? Are they not guardians of quality, stewards of our wine history, our landscapes and our terroirs, collaborators in our culinary culture, promoters of our regions' identity through the diversity of wines produced, and organizers of a virtuous partnership between humans and nature, blending tradition and innovation? In short, should their activity be included on the World Heritage List? The industry is perhaps more central than ever to profound debates that, through a range of approaches, we have attempted to elucidate.

To steer clear of oversimplified dualism in the quest for solutions, the various chapters in this book, as a result of research by scholars and practitioners, aim to capture the complexity of systems, the variety of profiles and paths, the necessity of a holistic and dynamic understanding of the issues and an infusion of insights extending beyond the wine sector or its traditional geographical boundaries.

While the solutions often require the involvement of a broad array of stakeholders, it is primarily the responsibility of those within the wine sector to demonstrate their ability to excel by addressing the negative externalities that may arise from their activities. It is their duty, along with their organizations and business partners, to undergo a likely deep transformation to minimize their ecological footprint, rediscover forgotten attributes, provide reassurance regarding health and quality of life and educate to foster new consumption habits acceptable to future generations. Only then can the finest aspects of viticulture be of service to consumers and our entire society.

To achieve these goals, the sector must evolve and continue to innovate for adaptation. It owns numerous resources, some of which are presented in this book, demonstrating how they can address four primary issues: financial enhancement, international expansion and competitiveness, communication and responsiveness to demand, sector organization and collective efforts. These issues, on their own, exemplify the complexity of forthcoming challenges and the array of involved stakeholders.

Technical advancements in the "vine-wine-commerce" sector should also be considered within the context of the ongoing digital transition: who will be the influencers for tomorrow's wines? Which communication tools and platforms will be used? What sources of information and knowledge about wines will be available? What purchasing methods and factors will be at play? What additional services will be offered beyond wine supply? These are all questions that could prompt wine companies to adjust their offerings and their approaches to wine production and sales.

Additional challenges loom on the horizon, intensifying the multifaceted uncertainties within the wine sectors. The first challenge revolves around the development of technical innovations, their adoption and, ultimately, the capacity of professional, political and research stakeholders to collaborate in addressing diverse issues. The second challenge pertains to the ongoing transformation of the winemaking profession, involving issues related to generational turnover and the introduction of new actors (financial, industrial or other agricultural players). This leads us to a third challenge concerning the future paths of farms, which may embrace to varying degrees multi-activity, diversification and agricultural multifunctionality.

These three challenges, combined with the constraints of societal and climate adaptation, have the potential to fundamentally alter the winemaking profession and, consequently, the structure of the sector.

Ultimately, these 16 chapters can be regarded as numerous avenues for contemplation and enhancement, for both industry professionals and academic researchers, to refine their craft and consequently align viticulture with market and societal expectations, as well as the demands of climate change. Even though wine industry participants are mindful of this, competitiveness in the forthcoming years will not be confined to the singular concern of production costs, despite its centrality within uncertain circumstances.

Agile, innovative and resilient enterprises will have the advantage. It is probable that only those companies that incorporate these diverse expectations and intricate challenges into their strategies and partnerships will be able to ensure the long-term sustainability of their activities. Therefore, it is reasonable to believe that the French wine sector, with its dynamism and innovative potential, possesses all of the qualities to confront the mentioned challenges and rejuvenate itself to provide products aligned with these expectations. This is the viticulture of the future!

List of Authors

Patrick AIGRAIN
FranceAgriMer
Montreuil
France

Paul AMADIEU
MRM
LabEx Entreprendre
Montpellier Management
Université de Montpellier
France

Clément ARNAL
Via Terra
Montpellier
France

Jérémy ARNAUD
Terroir Manager
Orange
Issy-les-Moulineaux
France

Yosr BEN TAHAR
PSB Research Lab
Paris School of Business
France

Rémi BRÉHONNET
CERIIM
Excelia Business School
La Rochelle
and
IRGO
Université de Bordeaux
France

Jean-Marie CARDEBAT
BSE
Université de Bordeaux
Pessac
and
INSEEC Grande École
OMNES Éducation
Bordeaux
France

Foued CHERIET
MoISA
Institut Agro Montpellier
France

Alvaro CUYA GAVILANO
IRGO
ECOr
Université de Bordeaux
France

Franck DUQUESNOIS
IRGO
ECOr
Université de Bordeaux
France

Jean-Philippe GALAN
IRGO
ECOr
Université de Bordeaux
France

Julien GRANATA
LabEx Entreprendre
Université de Montpellier
and
MIND
Montpellier Business School
France

Coralie HALLER
HuManiS
EM Strasbourg Business School
Université de Strasbourg
France

Hervé HANNIN
MoISA
Institut Agro Montpellier
France

Gérard HIRIGOYEN
IRGO
Université de Bordeaux
France

Théo JUSTY
MRM
Montpellier Management
Université de Montpellier
France

Eliane KARSAKLIAN
Université Sorbonne Nouvelle Paris 3
France
and
University of Illinois
Chicago
United States

Fanny LEPAGE
Université Laval
Quebec
Canada

Denis LESCOP
Montpellier Business School
France

Nada MAANINOU
HuManiS
EM Strasbourg Business School
Université de Strasboug
France

Carole MAUREL
MRM
Montpellier Management
Université de Montpellier
France

Sarah MUSSOL
MRM
Montpellier Management
Université de Montpellier
France

Mathieu NLEND-MOMNOUGUI
IRGO
ECOr
Université de Bordeaux
France

Iciar PAVEZ
MoISA
Institut Agro Montpellier
France

Estelle PELLEGRIN-BOUCHER
MRM
Montpellier Management
Université de Montpellier
France

Françoise PIERROT
MRM
Montpellier Management
Université de Montpellier
France

Louis-Antoine SAISSET
MoISA
Institut Agro Montpellier
France

Emmanuelle SAUVAGE
IRGO
ECOr
Université de Bordeaux
France

Jean-Marc TOUZARD
Innovation
INRAE
Montpellier
France

Jean-François TRINQUECOSTE
IRGO
Université de Bordeaux
France

Pascale WEBER
Willfreed SAS
Bordeaux
France

Index

A, B

C, D

E, F

G, H

I, K

L, M, N

P, R

S, T

U, V, W

Other titles from

iSTE

in

Agriculture, Food Science and Nutrition

2023

FLEURENCE Joël
Algae in Agrobiology: Realities and Perspectives

2020

CHERIET Foued, MAUREL Carole, AMADIEU Paul, HANNIN Hervé
Wine Management and Marketing: Opportunities for Companies and Challenges for the Industry

2019

YANNOU-LE BRIS Gwenola, SERHAN Hiam, DUCHAÎNE Sibylle, FERRANDI Jean-Marc, TRYSTRAM Gilles
Ecodesign and Ecoinnovation in the Food Industries

2016

JEANTET Romain, CROGUENNEC Thomas, SCHUCK Pierre, BRULÉ Gérard
Handbook of Food Science and Technology 1: Food Alteration and Food Quality
Handbook of Food Science and Technology 2: Food Process Engineering and Packaging
Handbook of Food Science and Technology 3: Food Biochemistry and Technology

Printed and bound by CPI Group (UK) Ltd, Croydon, CR0 4YY
16/09/2026
14963077-0007